Fundamental Toxicology for Chemists

Frontispiece *Potentially toxic and dangerous chemicals are now part of our everyday life, both in our homes and in our places of work.*
(Photo: Courtesy of H.G.J. Worth, The King's Mill Centre for Health Care Services, Sutton-in-Ashfield)

Fundamental Toxicology for Chemists

Edited by

John H. Duffus
Heriot-Watt University, Edinburgh

Howard G.J. Worth
King's Mill Centre for Health Care Services, Sutton-in-Ashfield

THE ROYAL
SOCIETY OF
CHEMISTRY
Information
Services

This textbook has been reviewed in detail by the IUPAC Commission of Toxicology and the IUPAC Committee on Chemistry and formally approved for publication at the IUPAC Congress at Guildford in August 1995.

ISBN 0–85404–529–5

Published by The Royal Society of Chemistry,
Thomas Graham House, Science Park, Milton Road, Cambridge CB4 4WF, UK

Typeset by Computape (Pickering) Ltd, Pickering, North Yorkshire, UK
Printed by Athenaeum Press Ltd, Gateshead, Tyne & Wear

Preface

There is, throughout the world, an increasing awareness of safety and a consciousness of the need for safety standards. Many countries have passed legislation concerned with safe practice in the work place. Although this has been led by the countries of Europe and North America and other developed countries, it is spreading and will spread to all areas of the world. This is exemplified by the progress that has been made in the United Kingdom. We merely mention this as an example because it is the situation with which we are most familiar.

The Health and Safety at Work Act, passed in 1974, required the appointment of safety representatives, the establishment of safety committees, and the inspection of the workplace by representatives and/or members of the safety committee. This has implications for the handling of chemicals where they are used in the workplace. More recently, regulations for the Control of Substances Hazardous to Health (COSHH) require laboratories and institutions to compile and retain inventories of chemicals used within their laboratories. The inventory must include statements concerning toxic properties, safe handling procedures, and actions that should be taken in the event of an accident or spillage. The safe handling of potentially toxic biological materials has been dealt with in the United Kingdom by the Department of Health and Social Security, which produced a series of documents culminating in the 'Code of Practice for the Prevention of Infection in Clinical Laboratories and Post Mortem Rooms'. This document was produced in 1978 and has now been superseded by 'Safe Working and the Prevention of Infection in Clinical Laboratories' which was produced by a Health and Safety Commission.

At the international level, the International Programme on Chemical Safety, a joint activity of the World Health Organization, the United Nations Environmental Programme, and the International Labour Organization, produces international chemical safety cards in conjunction with the Commission of the European Communities. This is only one example of what is happening through international bodies like the World Health Organization. As the input of such international bodies increases, so will national recognition throughout the world.

Much safety legislation and safe practice is concerned with the correct handling and use of chemicals. It is therefore reasonable that chemists should be aware of the dangers of the chemicals that are used within their laboratories, and that there should be documentation and legislation to help this safety process. But it is not confined to the laboratory. Chemicals are used increasingly in domestic and non-technical environments, where their safe handling is no longer solely the concern of qualified chemists. For instance, consider the use of domestic cleaners, solvents and detergents, weed killers and pesticides, and proprietary medicines. The question is therefore asked, who is the person to whom the public might turn to seek help and advice in the safe handling of chemicals? As like as not the answer that comes back is the chemist. It is not unreasonable that the chemist is seen as the person who can give help and advice on the handling of chemicals, on the toxic effects associated with them, and on how to deal with an incident when and if it occurs. However, this need is not recognized in curricula for the training of chemists, and indeed, apart from what they pick up incidentally as part of their educational progress, there is often no formal teaching of toxicology. This makes the chemist very vulnerable.

It is to address that problem that this book has been produced. It contains a proposed curriculum (Appendix 1) for the teaching of toxicology to chemists, which would give fundamental grounding of the basics, and for which the content of this book would be an adjunct.

The contents of this book have been carefully selected to cover those aspects of toxicology that are chemically related. The book clearly needs to include a wide range of biological and physical aspects from the effects of chemicals on target organs, to the effects on genetic replication resulting in mutagenicity and carcinogenicity, to the physical effects of radioactivity, to the biological effects on the environment, and to the physical behaviour of drugs in terms of their toxicokinetics and toxicodynamics. In addition, it includes consideration of the collection of data to allow meaningful toxic effects to be estimated and predicted. It has to be recognized that in this modern world of science and technology, if we are to reap the benefits of harnessing and handling chemicals, then this process itself results in a risk. Protocols for the safe handling of chemicals must therefore be assessed in this light, and there is a consequent need for risk management and assessment which is also dealt with. Finally, safe laboratory practices are considered within the context of all these other aspects.

An international team of authors has been invited to contribute to this book, all of whom have been selected for their recognized expertise in their chosen specialty and their interest in teaching and education.

The book has been produced under the auspices of the International Union of Pure and Applied Chemistry (IUPAC). Each chapter has been reviewed by the members of the IUPAC Commission on Toxicology and the Committee on the Teaching of Chemistry. We believe IUPAC is in a unique position to produce this book because of its access to worldwide expertise.

The book is completed by a glossary of terms used in toxicology which is abridged from the Glossary for Chemists of Terms Used in Toxicology (IUPAC Recommendations 1993), *Pure Appl. Chem.*, 1993, **65**, 2003–2122.

John H. Duffus
Howard G. J. Worth
(Editors)

Contents

Chapter 6   Risk Management
H. Paul A. Illing

Chapter 7   Exposure and Monitoring
Doug Templeton

Chapter 8   **Mutagenicity**
               Douglas McGregor

Chapter 9   **Carcinogenicity**
               Douglas McGregor

Chapter 10 **Reproductive Toxicity**
               Frank M. Sullivan

Chapter 11 Immunology and Immunotoxicology
John H. Duffus

Chapter 12   Skin Toxicity
Roger D. Aldridge

Chapter 13   Respiratory Toxicology
Raymond Agius

Chapter 18   **Ecotoxicity**
Martin Wilkinson

Chapter 19   **Radionuclides**
Milton Park

Acknowledgements

The editors would like to express their gratitude to all members of the IUPAC Commission on Toxicology and the Committee on the Teaching of Chemistry who made positive suggestions for improvement of the text, and their thanks to Mrs. Eileen Allin for the preparation of the manuscript.

JHD is deeply indebted to Dr. Alison L. Jones of the Scottish Poisons Information Bureau for checking the chapters on Neurotoxicity and Nephrotoxicity and suggesting valuable amendments, and also to Professor A. Dayan for compiling a glossary of terms used in immunotoxicology which provided the basis for the chapter on that subject.

Contributors

R. Agius, *Department of Public Health Sciences, University of Edinburgh, Medical School, Teviot Place, Edinburgh, EH8 9AG, UK*

R.D. Aldridge, *University Department of Dermatology, Royal Infirmary, Lauriston Buildings, Lauriston Place, Edinburgh, EH3 9YW, UK*

J.H. Duffus, *The Edinburgh Centre for Toxicology, Department of Biological Sciences, Heriot-Watt University, Riccarton, Edinburgh, EH14 4AS, UK*

J.S.L. Fowler, *Mulberry House, Magdalen Street, Eye, Suffolk, IP23 7AJ, UK*

B. Heinzow, *Untersuchungstelle für Umwelttoxikologie des Landes, Schleswig-Holstein, Fleckenstrasse 4, D-24105 Kiel 1, Germany*

F.R.M. Herber, *Coronel Laboratory for Occupational and Environmental Health, Free University of Amsterdam, Meibergdreef 15, 1105 AZ Amsterdam, The Netherlands*

H.P.A. Illing, *MRC Institute for Environment and Health, University of Leicester, PO Box 138, Lancaster Road, Leicester LE1 9HN, UK*

A.L. Jones, *Scottish Poisons Information Bureau, Royal Infirmary NHS Trust, Lauriston Place, Edinburgh, EH3 9YW, UK*

D. McGregor, *IARC, 150 Cours Albert Thomas, 69372 Lyon, Cedex 08, France*

M.V. Park, *The Edinburgh Centre for Toxicology, Department of Biological Sciences, Heriot-Watt University, Riccarton, Edinburgh, EH14 4AS, UK*

A.G. Renwick, *Clinical Pharmacology Group, University of Southampton, Medical and Biological Sciences Building, Bassett Crescent East, Southampton, SO16 7PX, UK*

F.M. Sullivan, *Division of Pharmacology and Toxicology, UMDS, St Thomas's Hospital, Lambeth Palace Road, London, SE1 7EH, UK*

D. M. Templeton, *Department of Clinical Biochemistry, Banting Institute, University of Toronto, Toronto, Canada, M5G 1L5*

M. Wilkinson, *Department of Biological Sciences, Heriot-Watt University, Riccarton, Edinburgh, EH14 4AS, UK*

G. Winneke, *Medizinisches Institut für Umwelthygiene, Heinrich-Heine-Universität, Düsseldorf, Auf'm Hennekamp 50, 40225 Düsseldorf, Germany*

H.G.J. Worth, *Clinical Chemistry Department, King's Mill Centre for Health Care Services, Mansfield Road, Sutton-in-Ashfield, NG17 4JL, UK*

CHAPTER 1

Introduction to Toxicology

JOHN H. DUFFUS

1.1 INTRODUCTION

Toxicology is the fundamental science of poisons. A poison is generally considered to be any substance that can cause severe injury or death as a result of a physicochemical interaction with living tissue. All substances are potential poisons since all of them can cause injury or death following excessive exposure. However, all chemicals can be used safely if exposure of people or susceptible organisms is kept below defined tolerable limits, *i.e.* if handled with appropriate precautions. If no tolerable limit can be defined, zero exposure methods must be used.

Exposure is a function of the amount (or concentration) of the chemical involved and the time of its interaction with people or organisms at risk. For very highly toxic substances, the tolerable exposure may be close to zero. In deciding what constitutes a tolerable exposure, the chief problem is often in deciding what constitutes an injury or adverse effect.

An adverse effect is defined as an abnormal, undesirable, or harmful change following exposure to the potentially toxic substance. The ultimate adverse effect is death but less severe adverse effects may include altered food consumption, altered body and organ weights, visible pathological changes, or simply altered enzyme levels. A statistically significant change from the normal state of the person at risk is not necessarily an adverse effect. The extent of the difference from normal, the consistency of the altered property, and the relation of the altered property to the total well-being of the person affected have to be considered.

An effect may be considered harmful if it causes functional or anatomical damage, irreversible change in homeostasis, or increased susceptibility to other chemical or biological stress, including infectious disease. The degree of harm of the effect can be influenced by the state of health of the organism. Reversible changes may also be harmful, but often they are essentially harmless. An effect which is not harmful is usually reversed when exposure to the potentially toxic chemical ceases. Adaptation of the exposed organism may occur so that it can live normally in spite of an irreversible effect.

1

In immune reactions leading to hypersensitivity or allergic responses, the first exposure to the causative agent may produce no adverse response, though it sensitizes the organism to respond adversely to future exposures.

The amount of exposure to a chemical required to produce injury varies over a very wide range depending on the chemical and the form in which it occurs. The extent of possible variation in harmful exposure levels is indicated in Table 1.1, which compares LD_{50} values for a number of potentially toxic chemicals. The LD_{50} value is more descriptively called the median lethal dose as defined below.

Table 1.1 *Approximate acute LD_{50} values for some potentially hazardous substances (Values obtained from the Merck Index, The Sigma–Aldrich Material Safety Data Sheets (Sigma–Aldrich Library of Chemical Safety Data), and Casarett and Doull's Toxicology)*

Substance	LD_{50} male rat (mg kg^{-1} body weight) Oral administration
Ethanol	7000
Sodium chloride	3000
Cupric sulfate	1500
DDT	100
Nicotine	60
Tetrodotoxin	0.02
Dioxin (TCDD)	0.02

DDT = 1,1,1-trichloro-2,2-bis(*p*-chlorophenyl)ethane; TCDD = 2,3,7,8-tetrachlorodibenzo-*p*-dioxin.

Median lethal dose (LD_{50}) is the statistically derived single dose of a chemical that can be expected to cause death in 50% of a given population of organisms under a defined set of experimental conditions. Where LD_{50} values are quoted for human beings, they are derived by extrapolation from studies with mammals or from observations following accidental or suicidal exposures.

The LD_{50} has often been used to classify and compare toxicity among chemicals but its value for this purpose is limited. A commonly used classification of this kind is shown in Table 1.2. Such a classification is arbitrary and not entirely satisfactory. For example, it is difficult to see why a substance with an LD_{50} of 200 mg kg^{-1} body weight should be regarded only as harmful while one with an LD_{50} of 199 mg kg^{-1} body weight is said to be toxic, when the difference in values cannot be statistically significant.

In decisions relating to chemical safety, the toxicity of a substance is less important than the risk associated with its use. Risk is the predicted or actual frequency (probability) of a chemical causing unacceptable harm or effects as a result of exposure of susceptible organisms or ecosystems. Assessment of risk is often assessment of the probability of exposure.

Conversely, safety is the practical certainty that injury will not result from exposure to a hazard under defined conditions; in other words, the high

Table 1.2 *An example of a classification of toxicity based on acute LD_{50} values (used in EC directives on classification, packaging, and labelling of chemicals)*

Category	LD_{50} orally to rat (mg kg^{-1} body weight)
Very toxic	Less than 25
Toxic	From 25 to 200
Harmful	From 200 to 2000

probability that injury will not result. Practical certainty is the numerically specified low risk or socially acceptable risk applied to decision making. For example, a chance of one in a million of suffering harm would generally be regarded as negligible and therefore safe.

In assessing permissible exposure conditions for chemicals, uncertainty factors are applied. An uncertainty factor is the mathematical expression of uncertainty that is used to protect populations from hazards which cannot be assessed with high precision. For example, the 1977 report of the US National Academy of Sciences Safe Drinking Water Committee proposed the following guidelines for selecting uncertainty (safety) factors to be used in conjunction with no observed effect level (NOEL) data. The NOEL should be divided by the following uncertainty factors:

1　An uncertainty factor of 10 should be used when valid human data based on chronic exposure are available.
2　An uncertainty factor of 100 should be used when human data are inconclusive, *e.g.* limited to acute exposure histories, or absent, but when reliable animal data are available for one or more species.
3　An uncertainty factor of 1000 should be used when no long-term, or acute human data are available and experimental animal data are scanty.

This approach is subjective and is being continually updated.

Safety control often involves the assessment of acceptable risk since total elimination of risk is often impossible. 'Acceptable' risk is the probability of suffering disease or injury that will be tolerated by an individual, group, or society. Assessment of risk depends on scientific data but its 'acceptability' is influenced by social, economic, and political factors, and by the perceived benefits arising from a chemical or process.

1.2　EXPOSURE TO POTENTIALLY TOXIC SUBSTANCES

Injury can be caused by chemicals only if they reach sensitive parts of a person or other living organism at a sufficiently high concentration and for a sufficient length of time. Thus, injury depends upon the physicochemical properties of the potentially toxic substances, the exact nature of the exposure circumstances, and the health and developmental state of the person or organism at risk.

Major routes of exposure are through the skin (topical), through the lungs (inhalation), or through the gastrointestinal tract (ingestion). In general, for exposure to any given concentration of a substance for a given time, inhalation is likely to cause more harm than ingestion which, in turn, will be more harmful than topical exposure.

1.2.1 Skin (Dermal or Percutaneous) Absorption

Many people do not realize that chemicals can penetrate healthy intact skin and so this fact should be emphasized. Among the chemicals that are absorbed through the skin are aniline, hydrogen cyanide, some steroid hormones, organic mercury compounds, nitrobenzene, organophosphate compounds, and phenol. Some chemicals, such as phenol, can be lethal if absorbed for a sufficient time from a fairly small area (a few square centimetres) of skin. If protective clothing is being worn, it must be remembered that absorption through the skin of any chemical which gets inside the clothing will be even faster.

1.2.2 Inhalation

Gases and vapours are easily inhaled but inhalation of particles depends upon their size and shape. The smaller the particle, the further into the respiratory tract it can go. Dusts with an effective aerodynamic diameter of between 0.5 and 7 μm (the respirable fraction) can persist in the alveoli and respiratory bronchioles after deposition there. Peak retention depends upon aerodynamic shape but seems to be mainly of those particles with an effective aerodynamic diameter of between 1 and 2 μm. Particles of effective aerodynamic diameter less than 1 μm tend to be breathed out again and do not persist in the alveoli or enter the gut (see below).

The effective aerodynamic diameter is defined as the diameter in micrometres of a spherical particle of unit density which falls at the same speed as the particle under consideration. Dusts of larger diameter either do not penetrate the lungs or lodge further up in the bronchioles and bronchi where cilia (the mucociliary clearance mechanism) can return them to the pharynx and from there to the oesophagus.

From the oesophagus, dusts are excreted through the gut in the normal way: it is possible that particles entering the gut in this way may cause poisoning as though they had been ingested in the food. A large proportion of dust breathed in will enter the gut directly and may affect the gut directly by reacting with it chemically or indirectly from contamination with micro-organisms. As already mentioned, some constituents of dust may be absorbed from the gut and cause systemic effects.

Physical irritation by dust particles or fibres can cause very serious adverse health effects but most effects depend upon the solids being dissolved. Special consideration should be given to asbestos fibres which may lodge in the lung and cause fibrosis and cancer even though they are insoluble and therefore not classical toxicants; similar care should also be taken with manmade mineral fibres. Insoluble particles may be taken in by the macrophage cells in the lung which normally remove invading bacteria. This is the process called phagocytosis. Phagocytosis is the process whereby certain body cells, notably macrophages and neutrophils engulf and destroy invading foreign particles. The cell membrane of the phagocytosing cell (phagocyte) invaginates to capture and engulf the particle. Hydrolytic enzymes are secreted round the particle to digest it and may leak from the phagocyte and cause local tissue destruction if the particle damages the phagocyte. If phagocytic cells are adversely affected by ingestion of insoluble particles, their ability to protect against infectious organisms may be reduced and infectious diseases may follow.

Some insoluble particles such as coal dust and silica dust will readily cause fibrosis of the lung. Others, such as asbestos, may also cause fibrosis depending on the exposure conditions.

Remember that tidal volume (the volume of air inspired and expired with each normal breath) increases with physical exertion; thus absorption of a chemical as a result of inhalation is directly related to the rate of physical work.

1.2.3 Ingestion

Airborne particles breathed through the mouth or cleared by the cilia of the lungs will be ingested. Otherwise, ingestion of potentially toxic substances in the work, domestic, or natural environment is likely to be accidental and common-sense precautions should minimize this. The nature of the absorption processes following ingestion is discussed elsewhere.

The importance of concentration and time of exposure has already been pointed out. It should be remembered that exposure may be continuous or repeated at intervals over a period of time; the consequences of different patterns of exposure to the same amount of a potentially toxic substance may vary considerably in their seriousness. In most cases, the consequences of continuous exposure to a given concentration of a chemical will be worse than those of intermittent exposures to the same concentration of the chemical at intervals separated by sufficient time to permit a degree of recovery. Repeated or continuous exposure to very small amounts of potentially toxic chemicals may be a matter for serious concern if either the chemical or its effects have a tendency to accumulate in the person or organism at risk.

A chemical may accumulate if absorption exceeds excretion; this may happen with substances that combine a fairly high degree of lipid solubility with stability.

1.3 ADVERSE EFFECTS

Adverse effects may be local or systemic. Local effects occur at the site of exposure of the organism to the potentially toxic substance. Corrosives and irritants always act locally.

Systemic effects occur at some distance from the site of exposure. Most substances which are not highly reactive are absorbed and distributed around the affected organism causing systemic injury at a target organ or tissue distinct from the absorption site. The target organ is not necessarily the organ of greatest accumulation. Adipose (fatty) tissue accumulates organochlorine pesticides to very high levels but does not appear to be harmed by them.

Some substances produce both local and systemic effects. For example, tetraethyl lead damages the skin on contact and is then absorbed and transported to the central nervous system where it causes further damage.

Effects of a chemical can accumulate even if the chemical itself does not. There is evidence that this may be true of the effects of organophosphate pesticides on the nervous system.

A particularly harmful effect that may accumulate is death of nerve cells, since nerve cells cannot be replaced, though damaged nerve fibres can be regenerated.

It will be clear that the balances between absorption and excretion of a potentially toxic substance and between injury produced and repair are the key factors in determining whether any injury follows exposure. All of the possible adverse effects cannot be discussed here but some aspects should be mentioned specifically.

Production of mutations, tumours and cancer, and defects of embryonic and fetal development are of particular concern.

Adverse effects related to allergies appear to be increasing. Allergy (hypersensitivity) is the name given to disease symptoms following exposure to a previously encountered substance (allergen) which would otherwise be classified as harmless. Essentially, an allergy is an adverse reaction of the altered immune system. The process, which leads to the disease response on subsequent exposure to the allergen, is called sensitization. Allergic reactions may be very severe and even fatal.

To produce an allergic reaction, most chemicals must act as haptens, *i.e.* combine with proteins to form antigens. Antigens entering the human body or produced within it cause the production of antibodies. Usually at least a week is needed before appreciable amounts of antibodies can be detected and further exposure to the allergen can produce disease symptoms. The most common symptoms are skin ailments such as dermatitis and urticaria, or eye problems such as conjunctivitis. The worst may be death resulting from anaphylactic shock.

Of particular importance in considering the safety of individuals is the possibility of idiosyncratic reactions. An idiosyncratic reaction is an excessive reactivity of an individual to a chemical, for example an extreme sensitivity to low doses as compared with an average member of the population. There is also the possibility of an abnormally low reactivity to high doses. An example of a group of people with an idiosyncrasy is the group that has a deficiency in the enzyme required to convert methaemoglobin (which cannot carry oxygen) back to haemoglobin; this group is exceptionally sensitive to chemicals like nitrites which produce methaemoglobin.

Another factor to be considered is whether the adverse effects produced by a potentially toxic chemical are likely to be immediate or delayed. Immediate effects appear rapidly after exposure to a chemical while delayed effects appear only after a considerable lapse of time.

Among the most serious delayed effects are cancers; carcinogenesis may take 20 or more years before tumours are seen in humans.

Perhaps the most difficult adverse effects to detect are those that follow years after exposure in the womb; a well established example of such an effect is the vaginal cancer produced in young women whose mothers have been exposed to diethylstilbestrol during pregnancy.

Another important aspect of adverse effects to be considered is whether they are reversible or irreversible. For the liver, which has a great capacity for regeneration, many adverse effects are reversible, and complete recovery can occur. For the central nervous system, in which regeneration of tissue is severely limited, most adverse effects leading to morphological changes are irreversible and recovery is, at best, limited. Carcinogenic and teratogenic effects may also be irreversible, but suitable treatment may reduce the severity of effects.

1.4   CHEMICAL INTERACTIONS

A major problem in assessing the likely effect of exposure to a chemical is that of assessing possible interactions. The simplest interaction is an additive effect: this is an effect which is the result of two or more chemicals acting together and which is the simple sum of their effects when acting independently.
In mathematical terms: $1 + 1 = 2, 1 + 5 = 6$ *etc.*
The effects of organophosphate pesticides are usually additive.

More complex is a synergistic (multiplicative) effect: this is an effect of two chemicals acting together which is greater than the simple sum of their effects when acting alone; it may be called synergism.
In mathematical terms: $1 + 1 = 4, 1 + 5 = 10$ *etc.*
Asbestos fibres and cigarette smoking act together to increase the risk of lung cancer by a factor of 40, taking it well beyond the risk associated with independent exposure to either of these agents.

Another possible form of interaction is potentiation. In potentiation, a substance which on its own causes no harm makes the effects of another chemical much worse. This may be considered to be a form of synergism.
In mathematical terms: $0 + 1 = 5, 0 + 5 = 20$ *etc.*
For example, isopropanol, at concentrations which are not harmful to the liver, increases (potentiates) the liver damage caused by a given concentration of carbon tetrachloride.

The opposite of synergism is antagonism: an antagonistic effect is the result of a chemical counteracting the adverse effect of another; in other words, the situation where exposure to two chemicals together has less effect than the simple sum of their independent effects. Such chemicals are said to show antagonism.
In mathematical terms: $1 + 1 = 0, 1 + 5 = 2$ *etc.*
For example, the toxicity of Cd(II) is greatly reduced by the simultaneous presence in food of Zn(II) and Ca(II).

1.5   TOLERANCE AND RESISTANCE

Tolerance is a decrease in sensitivity to a chemical following exposure to it or a structurally related substance. For example, cadmium causes tolerance to itself in some tissues by inducing the synthesis of the metal-binding protein, metallothionein. However, it should be noted that cadmium–metallothionein adheres in the kidney causing nephrotoxicity.

Resistance is almost complete insensitivity to a chemical. It usually reflects the metabolic capacity to inactivate and eliminate the chemical and its metabolites rapidly.

Genetic variation underlies both tolerance and resistance. For example, genetic variation in cytochrome P450 iso-enzymes explains variable resistance to carcinogenesis by polycyclic aromatic hydrocarbons. Similarly, inherited ability to acetylate aromatic amines provides resistance to the potential of beta-naphthylamine to cause bladder cancer.

1.6 TOXICITY TESTING

1.6.1 Dose–Response and Concentration–Response

The classic dose–response or concentration–response relationship is shown in Figure 1.1. This is a theoretical curve and in practice such a Gaussian curve is rarely found. This relationship forms the basis of the determination of the LC_{50} (the median lethal concentration) or the LD_{50} (the median lethal dose). The LC_{50} and LD_{50} are specific cases of the generalized values defined below.

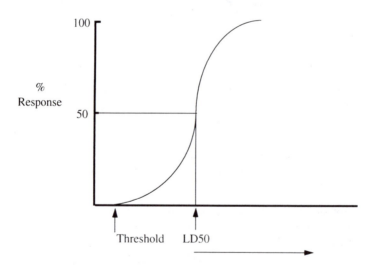

Figure 1.1 *The classic dose or concentration–response curve*

LC_n The exposure concentration of a toxicant lethal to n% of a test population.

LD_n The dose of a toxicant lethal to n% of a test population.

Median lethal concentration (LC_{50}) The statistically derived exposure concentration of a chemical that can be expected to cause death in 50% of a given population of organisms under a defined set of experimental conditions.

Median lethal dose (LD_{50}) The statistically derived single dose of a chemical that can be expected to cause death in 50% of a given population of organisms under a defined set of experimental conditions.

Another important value that may be derived from the relationship shown is the threshold dose or concentration, the minimum dose or concentration required to produce a detectable response in the test population. The threshold value can never be derived with absolute certainty and therefore the lowest observed effect level (LOEL) or the no observed effect level (NOEL) is used instead in deriving regulatory standards.

The use of the LD_{50} in the classification of potentially toxic chemicals has been described; it must be emphasized that such a classification is only a very rough guide to relative toxicity. The LD_{50} tells us nothing about sublethal toxicity. Any classification based on the LD_{50} is strictly valid only for the test population on which it is based and on the route of exposure. The LD_{50} tells us nothing about the shape of the dose–response curve on which it is based. Thus, two chemicals

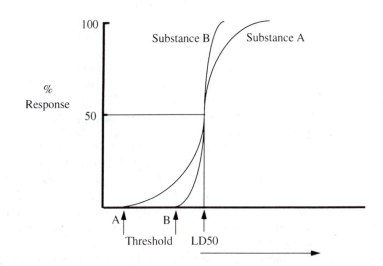

Figure 1.2 *Two substances with the same LD_{50} but different lower lethal thresholds*

may appear to be equally toxic since they have the same LD_{50} but one may have a much lower lethal threshold and kill members of an exposed population at concentrations where the other has no effect (see Figure 1.2.). Remember, these are theoretical curves and in practice Gaussian curves of this sort are rarely, if ever, found.

The determination and use of the LD_{50} is likely to decline in future as fixed dose testing becomes more widely used. In fixed dose testing, the test substance may be administered to rats or other test species at no more than 3 dose levels; the possible dose levels are preset legally to equate with a regulatory classification or ranking system. Dosing is followed by an observation period of 14 days. The dose at which toxic signs are detected is used to rank or classify the test materials.

A retrospective study of LD_{50} values showed that between 80 and 90% of those compounds which produced signs of toxicity but no deaths at dose levels of 5, 50 or 500 mg kg^{-1} body weight oral administration had LD_{50} values from the same studies of more than 25, from 25 to 200, or from 200 to 2000 mg kg^{-1} body weight. The European Economic Community classification has been based on this banding, namely very toxic, toxic, and harmful.

The initial test dose level should be selected with a view to identifying toxicity without mortality occurring. Thus, if a group of five male and five female rats is tested with an oral dose of 500 mg kg^{-1} body weight and no clear signs of toxicity appear, the substance should not be classified in any of the categories of toxicity applied.

If toxicity is seen but no mortality, the substance can be classed as 'harmful'. If mortality occurs, retesting with a dose of 50 mg kg^{-1} body weight is required. If no mortality occurs at the lower dose but signs of toxicity are detected, the substance would be classified as 'toxic'. If mortality occurs at the lower dose, retesting at 5 mg kg^{-1} body weight would be carried out and if signs of toxicity were detected and/or mortality occurred, the substance would be classified as 'very toxic'.

For full risk assessment, testing at 2000 mg kg^{-1} body weight is also required if no signs of toxicity are seen at 500 mg kg^{-1} body weight.

Fixed dose testing reduces the number of animals required and, because mortality need not occur, also greatly reduces any possible animal suffering. Fixed dose testing can also identify substances that have high LD_{50} values but still cause acute toxic effects at relatively low doses or exposures.

In assessing the significance of LD_{50} or other toxicological values, it is necessary to note the units used in expressing dosage. Normally dosage is expressed in mg kg^{-1} body weight but it may be expressed as mg cm^{-2} body surface area as this has been shown in a number of cases to permit more accurate extrapolation between animals of different sizes and from test species to humans.

For biocides, selective toxicity is the key property, so that they can be used to kill pests with minimal harm to other organisms. Selective toxicity depends upon differences in biological characteristics which may be either quantitative or qualitative. Hence, minimizing the amount of pesticide used and targeting its application is crucial to avoid harm to non-target organisms.

Toxicity testing is primarily aimed at establishing, by tests on laboratory animals, what effects chemicals are likely to have on human beings who may be exposed to them. On a body weight basis, it is assumed for toxicity data extrapolation that humans are usually about 10 times more sensitive than rodents. On a body surface-area basis, humans usually show about the same sensitivity as test mammals, *i.e.* they respond to about the same dose per unit of body surface area. Knowing the above relationships, it is possible to estimate the exposure to a chemical that humans should be able to tolerate.

In many countries there is now a defined set of tests that must be carried out on every new chemical that is to be used or produced in an appreciable quantity, usually above 1 tonne per year. Table 1.3 gives an example of test requirements applicable in a number of such countries.

Table 1.3 *Example of test requirements for new chemicals used for chemicals notification and hazard assessment in some countries*

Base set information

1 IDENTITY OF THE SUBSTANCE
1.1 Name
1.1.1 Names in the IUPAC nomenclature
1.1.2 Other names (usual name, trade name, abbreviation)
1.1.3 CAS number (if available)
1.2 Empirical and structural formula
1.3 Composition of the substance
1.3.1 Degree of purity (%)
1.3.2 Nature of impurities, including isomers and by-products
1.3.3 Percentage of (significant) main impurities
1.3.4 If the substance contains a stabilizing agent or an inhibitor or other additives, specify: nature, order of magnitude: ... ppm; ... %
1.3.5 Spectral data (UV, IR, NMR)
1.4 Methods of detection and determination
A full description of the methods used or the appropriate bibliographical references

2 INFORMATION ON THE SUBSTANCE
2.1 Proposed uses
2.1.1 Types of use
Describe: the function of the substance
the desired effects
2.1.2 Fields of application with approximate breakdown
(a) closed system
– industries
– farmers and skilled trades
– use by the public at large
(b) open system
– industries.
– farmers and skilled trades
– use by the public at large
2.2 Estimated production and/or imports for each of the anticipated uses or fields of application
2.2.1 Overall production and/or imports in order of tonnes per year 1, 10, 50, 100, 500, 1000, and 5000
– first 12 months
– thereafter
2.2.2 Production and/or imports, broken down in accordance with 2.1.1 and 2.1.2, expressed as a percentage
– first 12 months
– thereafter
2.3 Recommended methods and precautions concerning:
2.3.1 handling
2.3.2 storage
2.3.3 transport

Table 1.3 (*cont.*)

Base set information

2.3.4   fire (nature of combustion gases or pyrolysis, where proposed uses justify)
2.3.5   other dangers, particularly chemical reaction with water
2.4   **Emergency measures in the case of accidental spillage**
2.5   **Emergency measures in the case of injury to persons**
           (*e.g.* **poisoning**)

3   **PHYSICOCHEMICAL PROPERTIES OF THE SUBSTANCE**
3.1   **Melting point**
3.2   **Boiling point**
           °C at ... Pa
3.3   **Relative density ($D_4{}^{20}$)**
3.4   **Vapour pressure**
           Pa at ... °C
3.5   **Surface tension**
           $N\,m^{-1}$ (... °C)
3.6   **Water solubility**
           $mg\,l^{-1}$ (... °C)
3.7   **Fat solubility**
           Solvent–oil (to be specified)
           $mg\,100g^{-1}$ solvent (... °C)
3.8   **Partition coefficient**
           n-octanol/water
3.9   **Flash point**
           ... °C. Open cup and closed cup
3.10   **Flammability**
3.11   **Explosive properties**
3.12   **Auto-flammability**
           ... °C
3.13   **Oxidizing properties**

4   **TOXICOLOGICAL STUDIES**
4.1   **Acute toxicity**
4.1.1   Administered orally
           LD_{50} ($mg\,kg^{-1}$)
           Effects observed, including in the organs
4.1.2   Administered by inhalation
           LC_{50} (ppm)
           Duration of exposure in hours
           Effects observed, including in the organs
4.1.3   Administered cutaneously (percutaneous absorption)
           LD_{50} ($mg\,l^{-1}$)
           Effects observed, including in the organs
4.1.4   Substances other than gases shall be administered *via* two routes at least one of
           which should be the oral route. The other route will depend on the intended use
           and on the physical properties of the substance. Gases and volatile liquids should
           be administered by inhalation (a minimum period of administration of 4 hours).
           In all cases, observation of the animals should be carried out for at least 14 days.
           Unless there are contra-indications, the rat is the preferred species for oral and
           inhalation experiments. The experiments in 4.1.1, 4.1.2, and 4.1.3 shall be
           carried out on both male and female subjects.
4.1.5   Skin irritation

The substance should be applied to the shaved skin of an animal, preferably an albino rabbit.

Duration of exposure in hours

4.1.6 Eye irritation. The rabbit is the preferred animal.

Duration of exposure in hours

4.1.7 Skin sensitization. To be determined by a recognized method using a guinea-pig.

4.2 Sub-acute toxicity

4.2.1 Sub-acute toxicity (28 days)

Effects observed on the animal and organs according to the concentrations used, including clinical and laboratory investigations.

Dose for which no toxic effect is observed

4.2.2 A period of daily administration (5 to 7 days per week) for at least 4 weeks should be chosen. The route of administration should be the most appropriate having regard for the intended use, the acute toxicity, and the physical and chemical properties of the substance. Unless there are contra-indications, the rat is the preferred species for oral and inhalation experiments.

4.3 Other effects

4.3.1 Mutagenicity (including carcinogenic pre-screening test)

4.3.2 The substance should be examined during a series of two tests, one of which should be bacteriological, with and without metabolic activation, and one non-bacteriological.

5 ECOTOXICOLOGICAL STUDIES

5.1 Effects on organisms

5.1.1 Acute toxicity for fish

LC_{50} (ppm)

Duration of exposure

Species selected (one or more)

5.1.2 Acute toxicity for Daphnia

LC_{50} (ppm)

Duration of exposure

5.1.3 Acute toxicity for algae

LC_{50} (ppm)

Duration of exposure

Species selected (one or more)

5.2 Degradation: biotic and abiotic

The BOD and the BOD/COD ratio should be determined as a minimum

6 POSSIBILITY OF RENDERING THE SUBSTANCE HARMLESS

6.1 For industry/skilled trades

6.1.1 Possibility of recovery

6.1.2 Possibility of neutralization

6.1.3 Possibility of destruction:
  – controlled discharge
  – incineration
  – water purification station
  – others

6.2 For the public at large

6.2.1 Possibility of recovery

6.2.2 Possibility of neutralization

6.2.3 Possibility of destruction:
  – controlled discharge
  – incineration.
  – water purification station
  – others

1.7 EPIDEMIOLOGY AND HUMAN TOXICOLOGY

Epidemiology is the analysis of the distribution and determinants of health-related states or events in human populations and the application of this study to the control of health problems. It is the only ethical way to obtain data about the effects of chemicals other than drugs on human beings and hence to establish beyond doubt that toxicity to humans exists. The following are the main approaches that have been used in epidemiology.

1.7.1 Cohort Study

A cohort is a component of the population born during a particular period and identified by the period of birth so that its characteristics (such as causes of death and numbers still living) can be ascertained as it enters successive time and age periods. The term 'cohort' has broadened to describe any designated group of persons followed or traced over a period of time, for example as below in a cohort study (prospective study).

In a cohort study, one identifies cohorts of people who are, have been, or in the future may be exposed or not exposed, or exposed in different degrees, to a factor or factors hypothesized to influence the probability of occurrence of a given disease or other outcome. Alternative terms for such a study—follow-up, longitudinal, and prospective study—describe an essential feature of the method, observation of the population for a sufficient number of person-years to generate reliable incidence or mortality rates in the population subsets. This generally means studying a large population, study for a prolonged period (years), or both.

An incidence rate is the value obtained by dividing the number of people developing a disease in a defined period by the population at risk of developing that disease during this period, sometimes expressed as person-time. A mortality rate is the value obtained by dividing the number of persons dying during a given period by the total size of the population at risk of dying during this period.

Cohort studies involve cohort analysis. Cohort analysis is the tabulation and analysis of morbidity or mortality rates in relationship to the ages of the members of the cohort, identified by their birth period, and followed as they pass through different ages during part or all of their life span. In certain circumstances such as studies of migrant populations, cohort analysis may be performed according to duration of residence in a country rather than year of birth, in order to relate health or mortality experience to duration of exposure.

1.7.2 Retrospective Study

A retrospective study is used to test aetiological hypotheses in which inferences about exposure to the putative causal factor(s) are derived from data relating to characteristics of the persons or organisms under study or to events or experiences in their past: the essential feature is that some of the persons under study have the disease or other outcome condition of interest, and their characteristics and past experiences are compared with those of other, unaffected persons. Persons who differ in the severity of the disease may also be compared.

1.7.3. Case Control Study

A case control study starts with the identification of persons with the disease (or other outcome variable) of interest, and a suitable control (comparison, reference) group of persons without the disease. The relationship of an attribute to the disease is examined by comparing the diseased and non-diseased with regard to how frequently the attribute is present or, if quantitative, the levels of the attribute, in the two groups.

1.7.4 Cross-sectional Study (of Disease Prevalence and Associations)

A cross-sectional study examines the relationship between diseases (or other health-related characteristics) and other variables of interest as they exist in a defined population at one particular time. Disease prevalence rather than incidence is normally recorded in a cross-sectional study and the temporal sequence of cause and effect cannot necessarily be determined.

1.7.5 Confounding

Confounding is one of the biggest difficulties in carrying out a successful epidemiological investigation.

Confounding can occur in a number of different ways. Firstly, there is the situation in which the effects of two processes are not distinguishable from one another: this leads to the situation where distortion of the apparent effect of an exposure on risk is brought about by the association of other factors which can influence the outcome. Secondly, there is the possibility of a relationship between the effects of two or more causal factors as observed in a set of data, such that it is not logically possible to separate the contribution that any single causal factor has made to an effect. Finally, there is the situation in which a measure of the effect of an exposure on risk is distorted because of the association of exposure with other factor(s) that influence the outcome under study.

A confounding variable (confounder) is defined as a changing factor that can cause or prevent the outcome of interest, is not an intermediate variable, and is not associated with the factor under investigation. Such a variable must be controlled in order to obtain an undistorted estimate of the effect of the study factor on risk.

1.8 BIBLIOGRAPHY

B. Ballantyne, T. Marrs, and P. Turner, eds. 'General and Applied Toxicology', (2 volumes). Macmillan, London, 1993.

B.M. Francis, 'Toxic Substances in the Environment', John Wiley, New York, 1994.

C.D. Klaassen, ed. 'Casarett and Doull's Toxicology', McGraw-Hill, New York, 5th edn, 1996.

J.M. Last, 'A Dictionary of Epidemiology', Oxford University Press, 1988.

N.H. Stacey, ed. 'Occupational Toxicology', Taylor and Francis, London, 1993.

J.A. Timbrell, 'Introduction to Toxicology', Taylor and Francis, London, 2nd edn, 1995.

J.A. Timbrell, 'Principles of Biochemical Toxicology'. Taylor and Francis, London, 2nd edn, 1992.

CHAPTER 2

Toxicokinetics and Toxicodynamics - 1

ROBERT F. M. HERBER

2.1   INTRODUCTION

The father of toxicology, Philippus Aureolus Theophrastus Bombastus von Hohenheim (better known as Paracelsus), formulated early in 1538 the most important thesis in toxicology: 'What is it that is not a poison? All things are poisons and none that are not. Only the dose decides that a thing is not poisonous'.

Since all chemicals can produce injury or death under some exposure conditions, there is no such thing as a 'safe' chemical in the sense that it will be free of injurious effects under all circumstances of exposure. However, it is also true that there is no chemical that cannot be used safely by limiting the dose or exposure. Thus, reference should not be made to toxic and non-toxic substances or compounds, but rather to a toxic or non-toxic dose. A dose is the total amount of a substance administered to, taken, or absorbed by, an organism. A well-known example is sodium chloride, in small quantities necessary (essential) for human life, in larger quantities used as a flavour in foods and, in some parts of Asia, being used, in the past, as a suicide agent.

In food toxicology, substances are sometimes defined as essential or non-essential for humans. The distribution of the concentration of non-essential elements in different tissues (and blood) follows a non-Gaussian pattern, as these elements are not present unless there has been exposure to this non-essential compound. The normal concentration will be zero or close to zero. The concentration of essential elements, however, should be around the optimal concentration, and as a result the body keeps the concentration of such elements in equilibrium. This is called homeostasis.

Adverse or toxic effects in a biological system are not produced by a chemical agent unless that agent or its biotransformation products reaches appropriate sites in the body at a concentration and for a length of time sufficient to produce toxic effects. The toxic effect is thus dependent on the chemical and physical properties of the agent, the exposure situation, and the susceptibility of the cell, the biological system, or subject. Adverse effects may differ from undesirable to death, with all possible effects between, dependent on the compound and dose.

When a subject is exposed *via* the mouth (oral), the lungs, or other tissues, intake occurs. Following intake, a proportion of the compound (up to 100%) will be absorbed by the body. This is called the uptake. The difference between the intake and the uptake is the amount of the substance that leaves the body *via* the urine, faeces, exhaled air, hair, nails, sweat, *etc.* Compounds retained in the body will be metabolized or remain unchanged and be transported through the body. Toxic effects may appear during metabolism (*e.g.* in the liver), and by toxic (un)metabolized compounds in different organs. It is important to be aware of the possible different chemical composition (speciation) of a compound, and also the different effect of a compound when it is outside or inside the body. As an example, let us consider the element chromium. Chromium (III) is essential and is involved in glucose metabolism of the body. Higher concentrations of Cr(III) will not be absorbed, but are excreted (limited uptake). Cr(III) in higher concentrations, however, may cause dermal problems. Non-essential Cr(VI) compounds will be absorbed if they are water soluble and be metabolized to Cr(III) in the body. Thus a high concentration of Cr(III) may occur which will cause renal toxicity. Soluble Cr(VI) compounds also have an irritant and corrosive effect on the skin, eyes, and lungs. However, insoluble Cr(VI) cannot enter the body and if inhaled as an aerosol, remains in the lungs. Long-term exposure (at least 6 months) may result in bronchial cancer.

Thus, it will be clear that both the physical and the chemical properties may be responsible for toxicity.

2.2   TOXICOKINETICS

2.2.1   Biotransformation

The toxicity of a compound is dependent on the dose administered. Thus, higher toxicity is observed after a high dose of a compound than after a low dose (see Dose–Effect Relationships). The dose, however, is not the concentration of the administered compound in the body, but rather the concentration of the compound or metabolite in the target organ.

The concentration attained in the target organ depends on the deposition of the compound or metabolite. Deposition depends on its absorption (uptake), distribution, biotransformation, and excretion. The quantitation of the time course of absorption, distribution, biotransformation, and excretion is called toxicokinetics and is dealt with in Chapter 3.

Biotransformation may take place through the action of several enzymes, which may transform a lipid-soluble compound into water-soluble metabolites. There are two types of reactions: Phase I reactions, which involve oxidation, reduction, and protolysis, and Phase II reactions which consist of conjugation or synthetic reactions.

A prime function of a Phase I reaction is to add or expose functional groups (*e.g.* –COOH, –NH$_2$, –OH, –SH). These functional groups then allow the compound to undergo Phase II reactions. Phase II reactions will conjugate the foreign compound or Phase I-derived metabolites with an endogenous molecule.

These conjugates are normally added to endogenous products to promote their secretion or transfer across hepatic, renal, and intestinal membranes.

The enzymes or enzyme systems that catalyse the biotransformation of foreign compounds are located mainly in the liver. The liver receives all the blood that has perfused the splanchnic (bowel) area, which contains nutrients and other foreign substances. The liver can extract these substances readily from the blood, and chemically modify them before they are stored, secreted into the bile, or released into the general circulation.

Other organs can also biotransform foreign compounds but compared with the liver are limited by what chemicals they can handle. The biotransformation enzyme-containing cells are the parenchymal cells in the liver, the proximal tubular cells in the kidney, the Clara and type II cells in the lungs, the mucosa lining intestinal cells, the epithelial cells in the skin, and the seminiferous tubules and Sertoli's cells in the testes.

In the liver there are several enzymes responsible for biotransformation. Often both Phase I and Phase II enzymes can convert foreign compounds to forms that are more readily excreted. These kinds of enzymes are called detoxification enzymes. In several cases, however, the metabolites are more toxic than the parent compounds. This is particularly true for some chemical carcinogens, organophosphates, and some compounds that cause cell necrosis in the lung, liver, and kidney. Toxic metabolites may occur or, in other cases, highly reactive intermediates are formed during the biotransformation. The terms toxification or bioactivation are often used to indicate the enzymic formation of reactive intermediates. These intermediates are thought to initiate the events that ultimately result in cell death, chemically induced cancers, teratogenesis, and several toxic reactions.

Factors influencing biotransformation include species differences, sex, age, time of day, season, nutritional and disease status, route, dose, and enzyme inhibition or reduction.

2.2.2 Phase I Reaction Enzymes

The oxidative cytochrome P450 system is responsible for many biotransformations. The other important oxidative biotransformation system is the mixed-function amine oxygenase. Cytochrome P450 oxidative reactions involve, for example, aliphatic and aromatic hydroxylation, epoxidation, dealkylation, deamination, N-hydroxylation, sulfoxidation, desulfoxidation, and oxidative dehalogenation.

Reductive biotransformations of the cytochrome P450 system include azo reduction, aromatic nitro reduction, and reductive halogenation.

Mixed-function amine oxygenase is capable of oxidizing nucleophilic nitrogen and sulfur atoms. This flavin-containing monooxygenase competes with the cytochrome P450 system in the oxidation of amines. Known substrates for this enzyme are thiocarbamides and thioureas.

Epoxide hydrolyse catalyses the hydration of arene oxides and aliphatic epoxides to their corresponding 1,2-dihydrodiols.

Esterases have as preferred substrates aliphatic and aromatic esters, acetyl esters, and choline esters (the enzyme cholinesterase).

Aldehydes, alcohols, and ketones are often biotransformed by oxidation or reduction. The most important enzymes are aldehyde/ketones reductase, alcohol dehydrogenase, and aldehyde dehydrogenase.

2.2.3   Phase II Reaction Enzymes

Phase II enzymes require energy to drive their reactions. Glucuronosyltransferases catalyse glucuronidation which is a major conjugation reaction. The resulting glucuronides are excreted in the bile or urine. Examples of glucuronide conjugates are trichloroethanol, N-acetyl-N-phenyl hydroxylamine, 2-naphthylamine, thiophenol, and N,N'-diethyldithiocarbamic acid.

Examples of sulfotransferase conjugation substrates are phenols, catechols, hydroxylamines, and some primary and secondary alcohols.

N-Acetyl transferase substrates include aromatic primary amines, hydrazines, sulfonamides, and certain primary aliphatic amines.

Benzoates and phenyl acetate are amino acid substrates.

Glutathione-S-transferase substrates include methyl iodide, 3,4-dichloronitrobenzene, and benzoylchloride.

The enzyme rhodanase can detoxify cyanide by forming a thiocyanate.

2.3   TOXICODYNAMICS

2.3.1   Dose-effect relationships

In toxicology the important concept is the dose. It is also important to recognize the difference between dose and concentration. Dose is the concentration multiplied by time.

In the case of essential compounds there is a gradual transition from deficiency to effectiveness depending on the dose given. At low dose, when deficiency occurs, the metabolized or excreted amount of a substance is greater than the uptake, leading to a shortage in the cell or organ. A well-known example is vitamin deficiency, leading, in the case of vitamin C, to scurvy. An intake of vitamin C will redress the balance, and the scurvy will disappear. Thus, a certain minimum amount is necessary to prevent deficiency. This is illustrated in Figure 2.1, where the effect of an essential compound is shown as a function of dose. There is a gradual change of effects from deficiency to an essential level through to fatal toxicity.

Another example is zinc. There is a homeostatic control of zinc in blood plasma, but deficiency may lead to skin healing problems which are overcome by zinc treatment at the wound site. In excess, zinc may cause gastrointestinal problems if the dose is high enough (1 g or more).

The dose–effect relationship was originally developed by pharmacologists to study the relation between a drug and the effects or side effects on the body. These kinds of relationship can also be used in toxicology. They may be applied

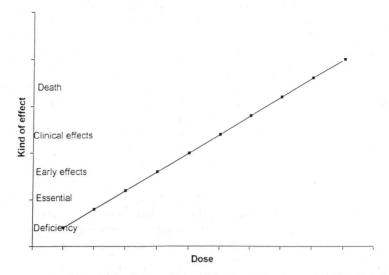

Figure 2.1 *A simple relationship between the dose and different effects for an essential compound. Note that with enhancing dose the seriousness of the effect increases*

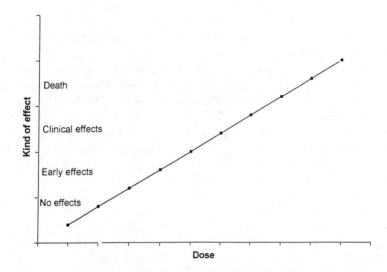

Figure 2.2 *A simple relationship between the dose and different effects for a non-essential compound*

to compounds as illustrated in Figure 2.2, which shows the relationship between the dose and different effects for a non-essential compound. The difference between Figures 2.1 and 2.2 is that, in the case of a non-essential compound, no benefit can be expected and thus sometimes such compounds are termed toxic.

Figures 2.1 and 2.2 refer to no effects, early effects, and clinical effects. The no effects area is clearly concerned with levels that have no effect. The maximum dose that produces no detectable change under defined conditions of exposure is referred to as the no effect level. It is not easy to determine whether an effect is detectable, and what the significance of the effect is. Thus a more practical measure is the no observed adverse effect level (NOAEL). An adverse effect is a change in morphology, physiology, growth, development, or life span of an organism that results in impairment of functional capacity or impairment of capacity to compensate for additional stress or increase in susceptibility to the harmful effects of other environmental influences. Early effects are not necessarily adverse effects and a NOAEL may be above the level causing an early effect but below the level causing a clinical effect.

Clinical effects are generally mentioned as adverse effects, *e.g.* occupational asthma (which may be caused by many compounds). Other examples of clinical effects are disturbances in the peripheral and central nervous system (*e.g.* tremor, polyneuropathy), and disturbances of liver and kidney function. As the body has only a few organs and the number of industrially used chemicals is about 70 000, very few early effects and clinical effects are specific for one compound. Thus, one effect may be caused by different compounds.

The branch of science dealing with the production of effects from the structure of a chemical compound is called quantitative structure–activity relationship (QSAR). QSAR has been used to predict toxic properties in the environment of organic fat-soluble compounds.

Dose–effect relationships are generally S-shaped, as can be seen in Figure 2.3. In this example there is an exponential rise above the baseline in urinary β_2-microglobulin, a parameter for assessing renal tubular damage, as a function of the dose of cadmium in urine. Beyond this, enhancing the dose gives only a moderate rise in the effect. The steepness of the dose–effect relationship determines the value at which the NOAEL must be set. The NOAEL has to be set at a lower dose when the curve rises steeply compared with a low rise.

Early effects are those that can be observed before adverse change has occurred. Examples are compounds that smell offensive at a no-effect level. Mercaptans, for example, are used as a warning odour in natural gas. At higher

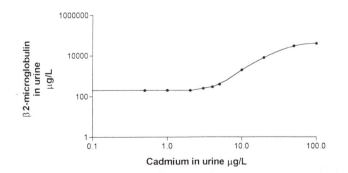

Figure 2.3 *The relationship between cadmium in urine (dose) and the appearance of β_2-microglobulin in urine (an early measure of renal tubule damage). Note the exponential rise after the horizontal part*

levels these are toxic. Thus in low concentrations a certain effect that is not adverse is used for a different purpose. Other examples of early effects are the excretion of certain enzymes and specific proteins into urine as a consequence of low exposure to certain compounds such as metals, solvents, pesticides, and drugs.

A dose–effect relationship exists for nearly all organ systems. But there are two exceptions, cancer and the immune system. These are dealt with in Chapters 9 and 11.

2.3.2   Biological Effect Monitoring

Biological effect monitoring (BEM) is defined as the continuous or repeated measurement and assessment of early biological effects in exposed humans, to evaluate ambient exposure and health risk by comparison with appropriate reference values based on knowledge of the probable relationship between ambient exposure and biological effects. BEM is used in environmental and occupational toxicology, and sometimes in forensic toxicology to monitor possible problems at an early stage. Mostly, BEM is carried out using readily available specimens such as urine, blood, and faeces.

When a BEM parameter is exceeded, the agent non-specific signs and symptoms are not necessarily related to exposure to specific agents, nor to conditions of work or environment. BEM parameters are sensitive, but non-selective, and thus may serve as a safety net when the exposure is not known (*e.g.* exposure to many different solvents or a mixture of solvents).

2.3.3   Dose-response Relationships

If in Figure 2.3 the effect is replaced by the response, a dose–response curve is obtained. A response may be the reaction of an organism or part of an organism (such as a muscle) to a stimulus, and as such is similar to a dose–effect curve. An alternative definition of response, is the proportion of a group of individuals that demonstrate a defined effect in a given time at a given dose rate. These kinds of dose–response relationships are quite different from the dose–effect relationships mentioned earlier. Dose–response (D–R) relationships in this sense can be used to differentiate between groups. If a D–R relationship for one group has a steeper slope than for another, the first group will be more sensitive to the dose than the latter. D–R relationships consequently can be used to detect sensitive groups.

2.3.4   Acute and Chronic Effects

As mentioned earlier, the dose is characterized by the integration of the concentration over time to produce a certain effect. This is useful in determining chronic effects, where the time scale is long, *e.g.* days or even years. It is also useful where there is a wide range of concentrations. Chronic effects are thus dependent on two parameters, time and concentration. It follows, therefore, that the effect will be chronic if the dose is persistent. Occasionally there may be a time shift between dose and effect, this being called a latent period. Generally, not only is a minimum dose needed to cause an effect (a dose above the NOAEL), but also a minimum time. This minimum time is called the minimum duration of exposure. In cases of chronic exposure, monitoring can take place by BEM or health surveillance, as a dose–effect (D–E) relationship exists (except for immunological effects).

In acute effects, the time component of the dose is not important, as the (high) concentration is responsible for the effect. Here a concentration–effect (or amount–effect) relationship may exist, but this is not generally true. Acute effects are almost always the cause of accidents (both in the home and at the workplace). The same is true for attempted poisoning (*e.g.* arsenic, thallium) and self-poisoning (suicide).

Other effects are local and systemic. Local effects are mainly to the skin, eyes, and respiratory tract, and it is often difficult to find a dose or concentration–effect relationship. With systemic effects, the whole body or many organs are affected and usually a D–E relationship will exist.

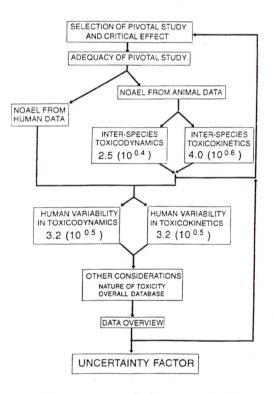

Figure 2.4 *From toxicity testing to health risk. See text for explanation*
(Reproduced with kind permission of the World Health Organization from 'Assessing Human Health Risks of Chemicals: Derivation of Guidance values for Health Based Exposure Limits', International Programme on Chemical Safety, World Health Organisation, Geneva, 1994)

2.4   TOXICITY TESTING AND HEALTH RISK

To test the toxicity of newly developed compounds a complex testing system has been developed. Figure 2.4 shows the steps involved in such a procedure. The first step is to select a critical effect of the compound. This is generally directed towards non-adverse effects, *e.g.* cancer, genotoxicity, teratogenicity, *etc.* The adequacy of such a study must then be considered. NOAELs from human and animal data should be assessed. Incorporating the interspecies variability in toxicokinetics and toxicodynamics can lead to new insights into both the human NOAELs and the pivotal study, and critical effect. This loop leads to human variability and, *via* other considerations, to a data overview. Sometimes the whole procedure will be performed several times. Finally, after the introduction of an uncertainty factor, a NOAEL is set.

In some countries a three-step procedure is developed. After the scientific process, pressure groups (mostly companies with unions, and the government) decide on a second step to determine if the draft NOAEL should be modified, and finally in an administrative step the NOAEL is published. It should be realized that such values only give the *risk* of effects, or consequently the *health risk.*

2.5   BIBLIOGRAPHY

G.D. Clayton and F.E. Clayton, 'Patty's industrial hygiene and toxicology', Wiley, New York, 1991 (10 volumes).

J.H. Duffus, S.S. Brown, N. de Fernicola, P. Grandjean, R.F. Herber, C.R. Morris, and J.A. Sokal, Glossary for chemists of terms used in toxicology, *Pure Appl. Chem.*, 1993, **65**, 2003–122.

'Environmental Health Criteria, International Programme on Chemical Safety', World Health Organization, Geneva (There are more than 170 volumes. Red books deal with compounds and yellow books with methods or techniques such as epidemiology, quality control, *etc.*)

C.D. Klaassen, ed. 'Casarett and Doull's Toxicology'. McGraw-Hill, New York, 5th edn., 1996.

CHAPTER 3

Toxicokinetics and Toxicodynamics - 2

A.G. RENWICK

3.1   INTRODUCTION

Toxicokinetics is the study of the movement of chemicals around the body. It includes absorption (transfer from the site of administration into the general circulation), distribution (*via* the general circulation into and out of the tissues), and elimination (from the general circulation by metabolism or excretion). The term toxicokinetics has useful connotations with respect to the high doses used in toxicity studies, but it may be misleading if interpreted as the 'movement of toxins around the body' since 'all things are toxic and it is only the dose which renders a compound toxic'.

Useful toxicokinetic data may be derived using a radiolabelled dose of the chemical, *i.e.* in which a proton in the molecule is replaced by a tritium atom or a carbon or sulfur atom is replaced by the radioactive equivalent (^{14}C or ^{35}S). Such studies are invaluable in following the fate of the chemical skeleton as it is transferred from the site of administration into the blood, is distributed to the tissues, and is eliminated as CO_2 or more likely as metabolites in air, urine, or bile. The advantage of using the radiolabelled chemical is that measured radioactivity reflects both the chemical and its metabolites, and this allows quantitative balance studies to be performed, *e.g.* to determine how much is absorbed, eliminated as CO_2 *etc.* However, such simple radioactive ADME (Absorption, Distribution, Metabolism, and Excretion) studies provide only a part of the total picture since the lack of chemical specificity in the methods does not allow an assessment of how much of the chemical is absorbed intact and how much is distributed around the body as the parent chemical. A further advantage of radiolabelled studies is that radiochromatographic methods can be invaluable in the separation and identification of metabolites, which is an important aspect of the fate of the chemical in the body. Thus, initial ADME studies define the overall fate of the chemical in the body and recognize the main chemicals (parent compound and/or metabolites) that are present in the circulation and in the urine and faeces following metabolism and excretion.

In recent years it been recognized that measurement of the circulating concentrations of the chemical and/or its metabolites can provide useful

26

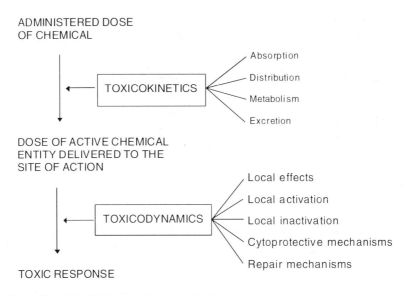

ADMINISTERED DOSE
OF CHEMICAL

TOXICOKINETICS

— Absorption

— Distribution

— Metabolism

— Excretion

DOSE OF ACTIVE CHEMICAL
ENTITY DELIVERED TO THE
SITE OF ACTION

TOXICODYNAMICS

Local effects

Local activation

Local inactivation

Cytoprotective mechanisms

Repair mechanisms

TOXIC RESPONSE

Figure 3.1 *The relationship between dose, toxicokinetics, and response*

information on the magnitude and duration of exposure of possible sites of toxicity. The term toxicokinetics is sometimes restricted to studies based on measurements of blood or plasma concentrations, since these provide a vital link between the dosing of experimental animals and the amounts of the chemical in the general circulation (Figure 3.1). Such information can be of great value in the interpretation of species differences in toxic response and the possible risk to humans of hazards identified in animal experiments. Also chemical-specific measurements are essential if the results of *in vitro* toxicity tests are to be interpreted logically. The ever increasing sensitivity of modern analytical techniques should allow the measurement of 'toxicokinetics' in humans receiving the compound at safe exposure levels. Thus, toxicokinetic differences between test animals and humans are open to direct measurement, and such data should increase confidence in the extrapolation process. Toxicokinetic data are also useful in extrapolating across different routes of exposure or administration as well as from single doses to chronic administration.

The toxicokinetics of a chemical are determined by measuring the concentrations of the chemical in plasma (usually) or blood at various times following a single dose. The fundamental parameters related to distribution and elimination are derived from data following an intravenous dose (Figure 3.2). The parameters relating to absorption from the gut, lungs *etc.* are derived from comparisons of data following the extravascular route of administration with an intravenous dose. Additional useful information can be obtained from measurements of the concentrations in plasma (or blood) over a period of 24 h in animals treated chronically with the chemical since the area under the plasma concentration-time curve often referred to as 'area under the curve' (AUC) is the best indication of exposure.

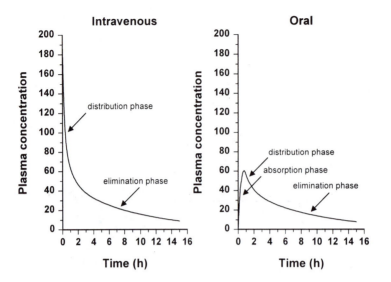

Figure 3.2 *The plasma concentration–time profiles following intravenous and oral administration*

The interpretation of toxicokinetic data requires an understanding of both the biological basis of the processes of absorption, distribution, and elimination and the way that simple measurements of plasma or blood concentrations can be converted into useful quantitative parameters which relate to these processes. The mathematics used to define and describe the movement of a chemical around the body can display various levels of sophistication and complexity. Compartmental analysis (Figure 3.3) allows the derivation of a mathematical equation which fits the data and allows the prediction of plasma concentrations outside the confines of the experiment. Physiologically based pharmacokinetic (PB-PK) modelling (Figure 3.4) allows a greater interpretation of the data in biologically relevant terms but requires a sophisticated database to produce valid results. This chapter will consider the biological basis of the processes of absorption, distribution, and elimination and describe the basic parameters, *e.g.* bioavailability, apparent volume of distribution, clearance, and half-life, which are most valuable because they are open to physiological interpretation. Each process can be considered in terms of the rate at which it occurs and the extent to which it occurs.

3.2   ABSORPTION

Strictly speaking the term absorption describes the process of the transfer of the parent chemical from the site of administration into the general circulation, and applies whenever the chemical is administered *via* an extravascular route (*i.e.* not by intravascular injection). 'Absorption' is also used to describe the extent to which a radiolabelled chemical is transferred from the site of administration into the excreta and/or expired air. However, many chemicals will be metabolized or transformed during their passage from the site of administration into the general circulation. Therefore, little parent chemical may reach the general circulation

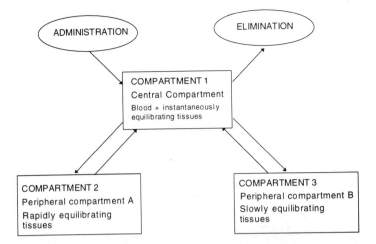

Figure 3.3 *Compartmental analysis. In the example shown the body is considered to consist of two peripheral compartments that equilibrate with the central compartment. The central compartment usually comprises blood and well-perfused tissues and essentially equilibrates instantaneously. In the example shown the chemical is eliminated from the central compartment. The number of compartments necessary in the model fitted to the data depends on the number of exponential terms necessary to describe the plasma concentration–time curve. The model can be used to estimate the concentration in the plasma or blood at any time after administration. The concentrations in the peripheral compartment(s) can also be estimated but these mathematical compartments are not equivalent to specific tissues or organs (The tissues and organs which make up each of the three compartments share only one common property, the rates of uptake of the chemical from blood)*

despite the fact that the radiolabel may all leave the site of administration and appear in the urine. This raises the possibility of confusion in discussing the extent of absorption depending on whether the data refer to the parent chemical *per se*, or to radiolabel (which will include the chemical plus metabolites). This confusion is resolved by the proper use of the term bioavailability described below.

3.2.1 Rate of Absorption

The rate of absorption may be important since it determines the peak plasma concentration and, therefore, the likelihood of acute toxic effects. Transfer of chemicals from the gut lumen, lungs, or skin into the general circulation involves movement across cell membranes: lipid-soluble molecules therefore tend to be absorbed more quickly than water-soluble ones. The gut wall and lungs provide a large and permeable surface area and allow rapid absorption; in contrast the skin is relatively impermeable and even highly lipid-soluble chemicals can enter only slowly. The rate of absorption depends on the extent of ionization of the chemical and, therefore, weak bases are not absorbed from the stomach because of its low pH, but are absorbed from the more alkaline intestinal contents. There are few membrane barriers to the absorption of subcutaneous and intramuscular doses and the absorption rate may be limited by the water solubility of the injected

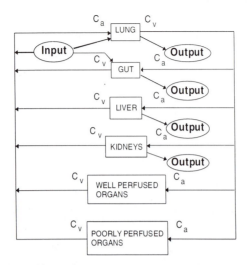

Figure 3.4 *Physiologically based pharmacokinetic model (PB-PK). The PB-PK model is derived from known rates of organ blood flow, the partition coefficient of the chemical between blood and tissue, and the rates of the processes of elimination (e.g. V_{max} and K_m for the enzymes). PB-PK modelling represents a powerful technique for estimating the dose delivered to specific tissues/organs and can facilitate inter-species extrapolation. Removal across an organ equals $(C_a - C_v)$ times blood flow*

material; slow absorption occurs with lipid-soluble compounds injected in an oily vehicle (which contrasts with the rapid digestion and absorption possible if such a dose is given *via* the gastrointestinal tract). Irrespective of the route of administration, the rate of absorption is determined from the early time points after dosing (Figure 3.5) and is usually described by an absorption rate constant or half-life.

3.2.2   Extent of Absorption

The extent of absorption is important in determining the total body exposure and, therefore, is an important variable during chronic toxicity studies and/or chronic human exposure. The extent of absorption depends on the extent to which the chemical is metabolized or broken down prior to reaching the general circulation and also on the rate of absorption compared with the rate of removal from the site of administration by other processes. Chemicals given *via* the gastrointestinal tract may be subject to a wide range of pH values and metabolizing enzymes in the gut lumen, gut wall, and liver before they reach the general circulation. The initial loss of chemical prior to it ever entering the blood is termed first-pass metabolism or pre-systemic metabolism; it may in some cases remove up to 100% of the administered dose so that none of the parent chemical reaches the general circulation. The other main reason for incomplete absorption of the parent chemical occurs when the rate of absorption is so slow that the chemical is lost from the body before absorption is complete. Examples of this include incomplete absorption of very water-soluble chemicals from the gut and their loss in the faeces, or incomplete dermal absorption, before the chemical is

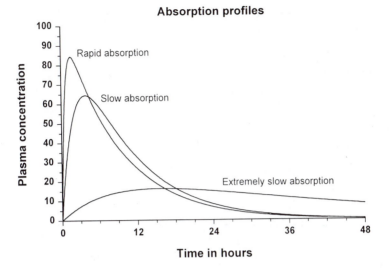

Figure 3.5 *The influence of the rate of absorption of a chemical on the plasma concentration–time curve. A relatively flat, low profile is obtained when the rate of elimination greatly exceeds the rate of absorption*

removed from the skin by washing. Whichever reason is responsible for the incomplete absorption of the chemical, it is essential that there is a parameter which defines the extent of transfer of the intact chemical from the site of administration into the general circulation. This parameter is the bioavailability, which is simply the fraction of the dose administered that reaches the general circulation as the parent compound. (The term bioavailability is perhaps the most misused of all kinetic parameters and is sometimes used incorrectly in a general sense as the relative amount available specifically to the site of toxicity.)

The fraction absorbed or bioavailability (F) is determined by comparison with intravenous (iv) dosing (where $F = 1$ by definition). The bioavailability can be determined from the area under the plasma concentration–time curve (AUC) of the parent compound (see Figure 3.6), or the % dose excreted in urine as the parent compound, *i.e.* for an oral dose

$$F = \frac{\text{AUC oral}}{\text{AUC iv}} \times \frac{\text{Dose iv}}{\text{Dose oral}}$$

$$F = \frac{\text{\% in urine as parent compound after oral dosing}}{\text{\% in urine as parent compound after intravenous dosing}}$$

3.3 DISTRIBUTION

Distribution is the reversible transfer of the chemical between the general circulation and the tissues. Irreversible processes such as excretion, metabolism, or covalent binding do not contribute to distribution parameters. The important distribution parameters relate to the rate and extent of distribution.

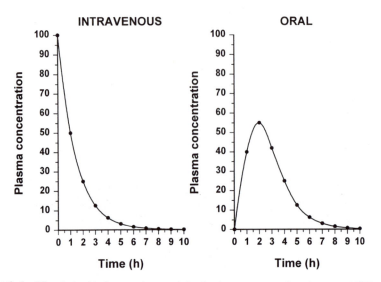

Figure 3.6 *The relationship between the area under the plasma concentration–time curve (AUC) and bioavailability. The bioavailability (fraction absorbed intact) is 1 for an intravenous dose. For other routes, the bioavailability is determined by the AUC for that route, divided by the AUC of an intravenous dose*

3.3.1   Rate of Distribution

The rate at which a chemical may enter or leave a tissue may be limited by two factors: (i) the ability of the compound to cross cell membranes and (ii) the blood flow to the tissues in which the chemical accumulates. The rate of distribution of highly water-soluble compounds may be slow due to their slow transfer from plasma into body tissues such as liver and muscle; highly water-soluble compounds do not accumulate in adipose tissue. In contrast very lipid-soluble chemicals can rapidly cross cell membranes, but the rate of distribution may be slow because they accumulate in adipose tissue and their overall distribution rate may be limited by blood flow to adipose tissue. Highly lipid-soluble chemicals may show two distribution phases: a rapid initial equilibration between blood and well perfused tissues, and a slower equilibration between blood and poorly perfused tissues (Figure 3.7). The rate of distribution is indicated by the distribution rate constant(s), which is determined from early time points after an intravenous dose. The rate constant(s) refers to a mean rate of removal from the circulation and may not correlate with uptake into a specific tissue (for which the PB-PK approach is more appropriate, see Figure 3.4). Once an equilibrium has been reached between the general circulation and a tissue, any process which lowers the blood (plasma) concentration will cause a parallel decrease in the tissue concentration (see Figure 3.8). Thus the elimination half-life measured from plasma or blood samples is also the elimination half-life from tissues.

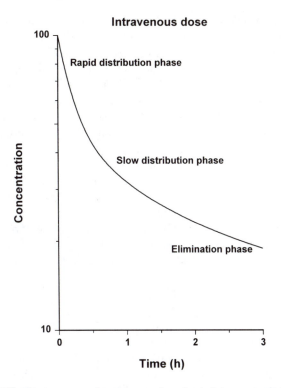

Figure 3.7 *The plasma concentration–time curve for a chemical that requires a three-compartment model (see Figure 3.3)*

3.3.2   Extent of Distribution

The extent of distribution of a chemical depends on the relative affinity of the blood or plasma compared with the tissues. Highly water-soluble compounds that are unable to cross cell membranes readily (*e.g.* tubocurarine) are largely restricted to extracellular fluid (about 13 litres per 70 kg body weight). Water-soluble compounds able to cross cell membranes (*e.g.* caffeine, ethanol) are largely present in total body water (about 41 litres per 70 kg body weight) unless a tissue(s) has a particular affinity for the chemical such as reversible tissue binding, when the blood (plasma) concentration will be lower than expected. Lipid-soluble compounds frequently show extensive uptake into tissues and may be present in the lipids of cell membranes, adipocytes, central nervous system *etc.*; the partitioning between circulating lipoproteins and tissue constituents is complex and may result in extremely low plasma concentrations. A further factor which may complicate the plasma:tissue partitioning is that some chemicals bind reversibly to circulating proteins such as albumin (for acid molecules) and α_1-acid glycoprotein (for basic molecules).

The internal environment of the brain is controlled by the endothelial cells of the blood capillaries to the brain which have tight junctions between adjacent cells, fewer and smaller pores, and little endocytosis. In consequence, water-soluble molecules cannot 'leak' into the brain between endothelial cells (as could happen, for example, in muscle capillaries) and are excluded from the brain. The endothelial membranes have specific transporters for the uptake of essential water-soluble nutrients and some ions and also for the exclusion of organic acids. This so-called blood-brain barrier serves to exclude most water-soluble com-pounds, so that central nervous system (CNS) toxicity may be limited. In contrast, lipid-soluble chemicals readily cross the blood-brain barrier and the CNS is a common site for toxicity (*e.g.* organic solvents). Similar permeability barriers are present in the choroid plexus, retina, and testes.

Although the extent and pattern of tissue distribution can be assessed by direct measurement of tissue concentrations in animals, comparable information is not possible from human studies and, therefore, the extent of distribution in humans has to be determined based solely on the concentrations remaining in plasma or blood after distribution is complete. The parameter used is the apparent volume of distribution (V) which relates the total amount of the chemical in the body (Ab) at any time after distribution is complete, to the circulating concentration (C).

$$V = \frac{Ab}{C}$$

V may be regarded as the volume of plasma in which the body load appears to have been dissolved and simply represents a dilution factor. The volumes of distribution of tubocurarine and caffeine are about 13 litres and 41 litres per 70 kg because of their restricted distribution. However, when a chemical shows a more extensive reversible uptake into one or more tissues then the plasma concentration will be lowered and the value of V will increase. For highly lipid-soluble chemicals, such as organochlorine pesticides, which accumulate in adipose tissue, the plasma concentration may be so low that the value of V may be many litres for each kg of body weight. Clearly this is an apparent and not a real volume but it is an important parameter since a high apparent volume of distribution is associated with a low elimination rate and a long half-life (see below). It must be emphasized that the apparent volume of distribution simply reflects the extent to which the chemical has moved out of the site of measure-ment (the general circulation) into tissues, but it does not reflect uptake into any specific tissue(s).

Information on the uptake into specific tissues requires sampling of that specific tissue, although PB-PK modelling can provide useful estimates. Once equilibrium has been reached for a tissue, the tissue:plasma ratio will remain constant, so that, as the chemical is eliminated from the plasma, the chemical will leave the tissue, maintaining the same ratio (Figure 3.8).

3.4   ELIMINATION

The parameter most commonly used to describe the rate of elimination of a chemical is the half-life (Figure 3.9). Most toxicokinetic processes are first-order reactions, *i.e.* the rate at which the process occurs is proportional to the amount

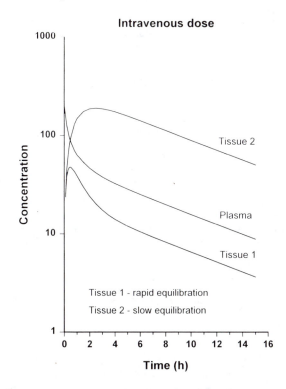

Figure 3.8 *Tissue distribution of a chemical after an intravenous dose. Tissue 1 shows a greater distribution rate than tissue 2 and therefore equilibrates more rapidly. Tissue 1 also shows a lower affinity than tissue 2 and therefore has a lower concentration after equilibration. The concentrations in plasma and both tissues decrease in parallel once an equilibrium has been established*

of chemical available. High rates occur at high concentrations and the rate decreases as the concentration decreases; in consequence the decrease is an exponential curve. The usual way to analyse exponential changes is to use logarithmically transformed data which converts an exponential into a straight line. The slope of the line is the rate constant (k) for the process and the half-life for the process is calculated as $0.693/k$. Therefore, rate constants and half-lives can be determined for absorption, distribution, and elimination processes.

There are two important variables that determine the rate at which a chemical can be eliminated from the body: (i) the clearance (which reflects the functional ability of the organs of elimination (*e.g.* liver or kidney) and (ii) the extent of distribution of the chemical from the general circulation into tissues.

The clearance of a chemical is determined by the ability of the organs of elimination to extract the chemical from the plasma or blood and permanently remove it by metabolism or excretion. (Note that this is different from distribution in which the chemical is free to leave the tissue and re-enter the blood when the concentration in the general circulation decreases.) The mechanisms of elimination depend on the chemical characteristics of the

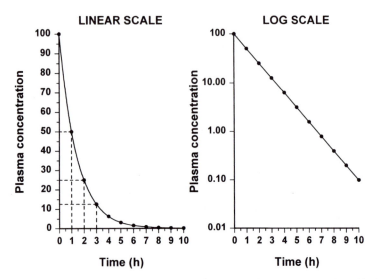

Figure 3.9   *The half-life of a chemical and its determination from plasma data. Logarithmic conversion allows the concentration data to be fitted by linear regression analysis; the half-life is calculated as 0.693/slope*

compound; therefore volatile chemicals are exhaled and water-soluble chemicals are eliminated in the urine and/or bile. Lipid-soluble chemicals are reabsorbed in the renal tubule and therefore are eliminated by metabolism to more water-soluble molecules (which are then eliminated in the urine and/or bile). Foreign compound metabolism is an enormous subject and involves a wide range of enzyme systems. Foreign chemicals (xenobiotics) may be metabolized by the enzymes of normal intermediary metabolism, *e.g.* esterases will hydrolyse ester groups. Alternatively chemicals may be metabolized by enzymes such as cytochrome P450, a primary function of which is xenobiotic metabolism. Species differences in metabolism can be a major source of differences in toxic response. The usual consequence of metabolism is the formation of an inactive excretory product so that species with low metabolizing ability will be likely to show greater toxicity. However, for many compounds metabolism is a critical step in the generation of a toxic or reactive chemical entity (bioactivation) and for such compounds high rates of metabolism will be linked with greater toxicity. If a chemical undergoes metabolic activation then toxicokinetic studies should measure both the parent chemical and the active metabolite. If the metabolite is so reactive that it does not leave the tissue in which it is produced (*e.g.* alkylating metabolites of chemical carcinogens) then toxicokinetic studies should define the delivery of the parent chemical to the tissues and the process of local activation should be regarded as part of tissue sensitivity (toxico-dynamics) since it is not strictly speaking part of toxicokinetics *i.e.* the movement of the chemical and/or metabolites around the body.

The best measure of the ability of the organs of elimination to remove the compound from the body is the clearance (CL)

$$CL = \frac{\text{rate of elimination}}{\text{plasma concentration}}$$

Since the rate of elimination is proportional to the concentration (see Figure 3.9) clearance is a constant for first-order processes and is independent of dose. It can be regarded as the volume of plasma (or blood) cleared of compound within a unit of time (*e.g.* ml min^{-1}).

Renal clearance depends on the extent of protein binding, tubular secretion, and passive reabsorption; it can be measured directly

$$CL = \frac{\text{rate of elimination in urine}}{\text{plasma concentration}}$$

The total clearance or plasma clearance (which is the sum of all elimination processes; renal + metabolic, *etc.*) is possibly the most important toxicokinetic parameter. It is measured from the total amount of compound available for removal (*i.e.* an intravenous dose) and the total area under the plasma concentration–time curve (AUC).

$$CL = \frac{\text{Dose iv}}{\text{AUC iv}}$$

Plasma clearance reflects the overall ability of the body to remove permanently the chemical from the plasma. Plasma clearance is the parameter that is altered by factors such as enzyme induction, liver disease, kidney disease, inter-individual or inter-species differences in hepatic enzymes or in some cases organ blood. The same clearance value applies once the chemical is in the general circulation, irrespective of the route of delivery. However, the bioavailability (F) will determine the proportion of the dose reaching the general circulation. Therefore, bioavailability has to be taken into account if clearance is calculated from data from a non-intravenous route (*e.g.* oral).

$$CL = \frac{\text{Dose oral} \times F}{\text{AUC oral}}$$

Measurement of Dose/AUC for an oral dose determines CL/F, which contains two potentially independent variables.

The overall rate of elimination, as indicated by the terminal half-life ($t_{\frac{1}{2}}$), is dependent on two physiologically related and independent variables:

$$t_{\frac{1}{2}} = \frac{0.693 \times V}{CL}$$

CL – the ability to extract and remove irreversibly the compound from the general circulation.

V – the extent to which the compound has left the general circulation in a reversible equilibrium with tissues

Therefore, a chemical may have a long half-life because the organs of elimination have a low ability to remove it from plasma and/or because it is extensively distributed to body tissue and little remains in plasma and is available for elimination. Chemicals that are extremely lipid-soluble and are sequestered in adipose tissue are eliminated slowly. Lipid soluble organochlorine compounds, which are not substrates for P450 oxidation due to the blocking effects of the chloro-substituents, are eliminated extremely slowly: for example, the half-life of 2,3,7,8-tetrachlorodibenzodioxin (TCDD) is about 8 years in humans.

3.5   CHRONIC ADMINISTRATION

Most toxicity studies involve continuous, or chronic administration of the chemical either *via* incorporation into the diet or by daily gavage doses. The kinetic parameters of a single dose (as discussed above) apply to chronic administration, but the exposure has to allow for the fact that not all of the previous dose(s) may have been eliminated when the subsequent dose is given. Therefore, there may be an increase in plasma concentration (and body load) until an equilibrium is reached in which the rate of elimination balances out the rate of input (Figure 3.10). The description of clearance given above can be rewritten as:

$$\text{Rate of elimination } (\mu g \, d^{-1}) = CL \, (\text{ml } d^{-1}) \times \text{plasma concentration } (\mu g \, \text{ml}^{-1})$$

i.e. the rate of elimination is proportional to plasma concentration. When doses of chemical are given daily the low plasma concentrations from the first dose may give a low rate of elimination such that not all of the chemical would be eliminated before the next dose is given. In consequence, the next dose will give

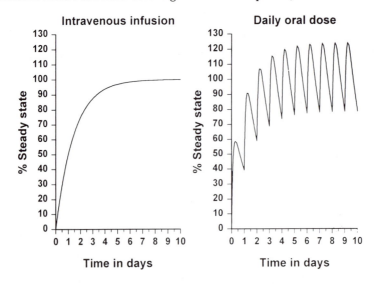

Figure 3.10 *The increase to steady-state plasma concentration following both a continuous intravenous infusion and a once-daily oral administration (e.g. daily in the diet)*

higher concentrations (due to carryover from the first dose) and, therefore, the rate of elimination will be higher. In consequence a greater proportion of the daily dose will be eliminated on the second day. The plasma concentrations and rates of elimination will continue to increase each day until the plasma concentrations are such that the daily dose is eliminated each day (Figure 3.10), *i.e.* an equilibrium or steady state is reached. At equilibrium the balance between input and output can be written as:

$$\text{(Input)}\frac{\text{Dose} \times F}{\text{Dose interval}} = CL \times C_{SS}\text{(Output)}$$

C_{SS} is the average steady-state plasma concentration which can be calculated by:

$$C_{SS} = \frac{\text{Dose} \times F}{CL \times \text{dose interval}}$$

It is important to realize that clearance is the same value throughout the buildup to steady state. Once steady-state has been reached:

$$CL = \frac{\text{Dose} \times F}{\text{AUC for dose interval}}$$

The equations above assume that CL is constant; the assumption is not correct if the chemical induces or inhibits its own elimination since clearance would be increased or decreased respectively after a period of chronic intake. The possibility that metabolism or excretion is saturated at the higher plasma concentrations during chronic intake is discussed below.

A further important toxicokinetic variable to be considered in the design and interpretation of chronic studies is the time taken to reach steady state. The extent of toxicity is usually proportional to the dose or the body load and the body load is given by the plasma concentration (at any time) multiplied by the apparent volume of distribution (which is constant), *i.e.* body load (Ab) equals $C \times V$. Since C increases with chronic intake until an equilibrium is reached (Figure 3.10) so Ab also increases to reach a steady state. The time taken to reach steady state is 4–5 times the elimination half-life and, therefore, the true duration of steady-state exposure in a toxicity study is the study duration minus 4–5 half-lives of the chemical. This is particularly important for chemicals that have a very long half-life; even in rodents the steady-state body load of TCDD will not be reached until after about 4 months of continuous treatment.

3.6 SATURATION KINETICS

The parameters described above are all independent of dose at low doses. However, at high doses and/or during chronic studies it is possible to overload or saturate any compound–protein interaction. Under such circumstances any increase in the concentration of the compound cannot give an increased rate of the process. The rate is at the maximum possible and is essentially independent of concentration.

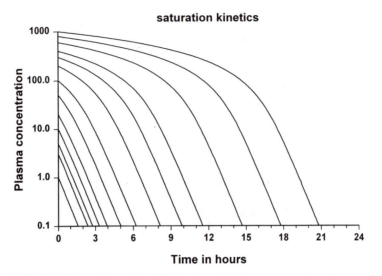

Figure 3.11 *The influence of saturation of elimination on the plasma concentration–time curve. The K_m of the elimination process was set at 100 μg ml^{-1} and the elimination is essentially log-linear once the concentrations have decreased to this level (i.e. first order). There is a slower decrease in concentration at higher concentrations because the enzyme is saturated (i.e. zero order)*

In simple mathematical terms this means that the reaction changes from first order to zero order. This is best described by Michaelis–Menten kinetics, *i.e.*

$$\text{Rate} = \frac{V_{max}C}{K_m + C}$$

At low concentrations C is less than the Michaelis constant K_m and, therefore, $(K_m + C)$ approximates to K_m. At such low concentrations the rate equals $(V_{max}/K_m) \times C$ and since V_{max} and K_m are constants the rate is proportional to the concentration (first order).

At high concentrations C is more than K_m and, therefore, $(K_m + C)$ approximates to C. At such high concentration the rate equals $(V_{max}/C) \times C$, *i.e.* V_{max} and therefore is a fixed maximum rate (zero order).

The consequences of this for a plasma concentration–time curve are shown in Figure 3.11. It is important to note that the terminal half-life is always determined at low concentrations and therefore is a first-order constant which does not show saturation kinetics. The best parameter to reflect saturation kinetics is the *CL* which is based on the total AUC and which includes the slower zero-order elimination phase. A classic example of this type of data is shown in the work of Dietz *et al.* on the solvent dioxane, in which saturation of metabolism resulted in a change of clearance in rats from 13.3 ml min^{-1} at 3 mg kg^{-1} to 1.0 ml min^{-1} at 1000 mg kg^{-1}. The renal tubular secretion process can also be saturated, as demonstrated for cyclohexylamine in rats by Roberts and Renwick, which resulted in a non-linear accumulation of the compound in the testes of rats (but not mice) which correlated with the dose–response for the testicular toxicity.

Important possible consequences of saturation of metabolism or excretion are that the chemical will accumulate to higher concentrations and that some normally minor alternative routes of elimination may become involved. The toxic effects seen at saturating doses may be of little or no relevance to lower non-saturating doses when the alternative route is a different pathway of metabolism which results in bioactivation of the chemical.

3.7 CONCLUSIONS

A common criticism of animal experiments is that they are 'not relevant' to humans. This is not true as a generalization, but there are instances where animal data are not relevant to human risk assessment due to the nature of the target organ for toxicity or due to the fate of the chemical in the body. Even when the target organ and pathways of metabolism are similar, inter-species differences in the fate of a chemical in the body complicate the interpretation of animal data in relation to human risk assessment. Information on the toxicokinetics of a chemical provides an understanding of the extent of absorption and distribution and the pathways and rates of elimination. Such data provide a vital link between animal experiments and human safety.

3.8 BIBLIOGRAPHY

J.D. de Bethizy and J.R. Hayes, 'Metabolism. A determinant of toxicity', in 'Principles and Methods of Toxicology', ed. A. Wallace Hayes. Raven Press, New York, 3rd edn, 1994, pp. 59–100.

F.K. Dietz, W.T. Stott, and J.C. Ramsey, Non-linear pharmacokinetics and their impact on toxicology: illustrated by dioxane, *Drug Metab. Rev.*, 1982, **13**, 963–81.

E.R. Garrett, 'Toxicokinetics', in 'Toxic Substances and Human Risk', eds. R.G. Tardiff and J.V. Rodricks. Plenum, London, 1987, pp. 153–237.

K. Krishnan and M.E. Andersen, 'Physiologically based pharmacokinetic modeling in toxicology', in 'Principles and Methods of Toxicology', ed. A. Wallace Hayes. Raven Press, New York, 3rd edn, 1994, pp. 149–188.

E.J. O'Flaherty, 'Differences in metabolism at different dose levels', in 'Toxicological Risk Assessment', eds. D.B. Clayson, D. Krewski, and I. Munro. CRC Press, Boca Raton FL, 1985, vol. 1, pp. 53–91.

A.G. Renwick, Data-derived safety factors for the evaluation of food additives and environmental contaminants, *Food Addit. Contam.*, 1993, **10**, 275–305.

A.G. Renwick, 'Toxicokinetics', in 'General and Applied Toxicology', eds. B. Ballantyne, T. Marrs, and P. Turner (2 volumes). Macmillan, London, 1993, pp. 121–51.

A.G. Renwick, 'Toxicokinetics – Pharmacokinetics in toxicology', in 'Principles and Methods of Toxicology', ed. A. Wallace Hayes. Raven Press, New York, 3rd edn, 1994, pp. 101–147.

A. Roberts and A.G. Renwick, The pharmacokinetics and tissue concentrations of cyclohexylamine in rats and mice, *Toxicol. Appl. Pharmacol.*, 1989, **98**, 230–42.

J.A. Timbrell, 'Biotransformation of xenobiotics', in 'General and Applied Toxicology', eds. B. Ballantyne, T. Marrs, and P. Turner. Macmillan, London, 1993, pp. 89–119.

WHO, 'Principles of Toxicokinetic Studies', Environmental Health Criteria 57. World Health Organisation, Geneva, 1986.

J.R. Withey, 'Pharmacokinetic differences between species', in 'Toxicological Risk Assessment', eds., D.B. Clayson, D. Krewski, and I. Munro. CRC Press, Boca Raton, FL, 1985, vol. 1, pp. 41–52.

CHAPTER 4

Data Interpretation

JOHN FOWLER

4.1 TYPES AND USES OF DATA

Data interpretation in toxicology is complicated not so much by the techniques that are involved, but rather because of the varied types and sources of data that must be addressed.

Potential hazard exists within every chemical, although the chances of this becoming important may be only slight, if opportunities for exposure to the chemical are limited.

Calculation of the chances of a potential hazard becoming a real threat is 'risk analysis'. The need to undertake risk analysis is central to the science of toxicology and is achieved by collection and interpretation of various data. For example, before risk analysis of a chemical in use can begin, data must be interpreted that pertain to:

– the intrinsic toxicological hazard possessed by the chemical
– the duration of contact of potential targets with the chemical
– the amount of chemical achieving an interaction with the target
– the inherent susceptibility of the target to the chemical.

Unless these data exist it is impossible to calculate the risk to the target that is attributable to the chemical.

4.2 GENERATION, STORAGE, AND RETRIEVAL OF DATA

Data are generated during surveys, testing, studies, and experiments and are accumulated in databases or databanks. Many of these databases or databanks are accessible to toxicologists: the majority are located in the USA, although there are several in the European Union and others in Canada and Switzerland.

The first stage of data interpretation is data retrieval, although before even this can commence it is necessary to know which data are required, and for what purpose they will be needed.

4.3   DATA RETRIEVAL

4.3.1   Information That May be Required

Typical headings of information that would correspond to the data profile structure of the International Register of Potentially Toxic Chemicals (IRPTC) are as follows: identifiers, properties, classification; production trade; production process; use; pathways into the environment; concentrations; environmental fate tests; environmental fate; chemobiokinetics; mammalian toxicity; special toxicity studies; effects on organisms in the environment; sampling/ preparation/analysis; spills; treatment of poisoning; waste management.

To obtain data under appropriate headings, a library must be identified which contains or has access to the materials to be consulted, *e.g.* to books, journals, abstracts and indexes. The Merck Index is widely available and is a particularly useful source; Current Contents, which is now available on floppy disc, allows a retrieval of full titles and authors of current papers; Chemical Abstracts, the largest index, provides summaries with citations from journals, patents, reports, specialist books, and conference proceedings, and there is Science Citation Index, which allows identification of related papers. Access online to computer-based databases or databanks may simplify and speed data retrieval: databanks contain pre-selected information in summary form; databases provide access without pre-selection or evaluation. Two important databanks are the United States Registry of Toxic Effects of Chemical Substances (RTECS) and the European Community's Environmental Chemicals Data and Information Network (ECDIN); examples of databases are TOXLINE, MEDLINE, and CANCERLINE which are free text databases.

4.3.2   Some Types of Data

Not all data arise from similar systems; for example the complex non-linear systems such as thermodynamics and meteorology do not behave predictably and a simple cause–effect relationship may not be assumed. Interpretation of data from such systems necessitates definition of the exact state of all the forces and matter involved, together with accurate measurement of all the interacting factors, in which case behaviour may be predictable at a certain level of probability. A pattern will emerge, but will never repeat itself exactly: this is the basis of the recently described Chaos Theory. According to this theory, health is regarded as chaotic, and in health there is an ability to respond to a large variety of adverse stimuli. On the other hand, disease is interpreted as a loss of flexibility or periodicity, and in disease there is lessened ability to adapt and respond to external stimuli.

The much older concept of cause-and-effect is seen to operate in the more simple linear systems and presently still forms the basis for most of our data interpretation in toxicology. The method of logical deduction is attributed to the Frenchman Descartes (1596–1650), who did most of his work in bed, whereas the idea of inductive reasoning and experimentation is associated with Francis

Bacon (1561–1626), an Englishman who died of the cold he caught whilst he tried the experiment of stuffing a chicken with snow. The inductive approach, which is based on observation and collection of data, is familiar to all scientists, having been adopted very widely during the 19th century [*e.g.* by Darwin] and the experimentation approach, is also well known, since it is currently fashionable and is utilised by nearly all present-day scientists.

4.3.3 Data Arising from the Study of Chemicals

A common plan when studying a chemical is to undertake a carefully controlled experiment preferably in the target species. If the chemical is a candidate pharmaceutical agent, it follows that the definitive data will arise from studies in man. Prior to studies in man, or where the target is less well defined (for example for industrial or agricultural applications), it is the present practice to employ surrogates, including *in vivo* and *in vitro* methods, for the presumed target species or system.

In a typical plan, when seeking to discover the biological properties of an unknown chemical, quantitative data are gathered regarding various predicted and expected effects, and additionally a search is made for other actions that were not expected. Unexpected data from the latter (screening) studies tend to be qualitative rather than quantitative in the first place and interpretation of qualitative data is usually subjective, being based on experience and intuitive reasoning.

It is usually necessary, therefore, that qualitative data of interest or of concern, which have been generated from screening approaches, must be studied further and in greater depth using quantitative techniques. Interpretation of quantitative data will be by objective methods.

4.4 HANDLING QUANTITATIVE DATA USING STATISTICAL ANALYSIS

Statistical treatment of data will enable extraction of important features; for example indicators of central tendency such as the mean, mode, or median and indicators of spread about the centre, such as the standard deviation and interquartile range.

4.4.1 Null Hypothesis

In chemical testing, it is usual to set up a study-based statistical model in order to provide data suitable for statistical analysis. For example a chemical under study (the test substance) is usually applied to a model system whilst comparing its actions (if any) with those arising from simultaneous application of a bland or otherwise relevant reference or standard substance (the control). Often called the 'sham-treated group' this controlled group may be exposed to the solvent-vehicle used to solubilize the test substance (for example corn oil or water depending on the solubility of the test material) and will be identical in every way with the test group, except that it will not receive the test substance. At the outset of such a study it is assumed that there will be no difference between the 'test' and 'sham-treated' groups (null hypothesis). Data are collected from each group to enable comparisons to be made. If subsequently data from the groups are found to differ, then it is assumed that this difference arose due to prior treatment with the test substance, *i.e.* it is attributable to exposure to the chemical under investigation.

The possibility or probability that this difference is indeed attributable to an effect of the chemical may be calculated using statistical techniques. Application of such techniques will allow determination of the confidence that should be attached to the results. The process of data interpretation using statistics may be illustrated by the decision-tree approach, which is illustrated in Figure 4.1.

4.4.2   Generation of Data Relating to Chemical Safety

The data needed for risk estimation are currently provided by toxicological investigations that depend mainly on laboratory experiments. For a summary of the types of non-biological data and biological-effect data that are needed in order to make assessments of biological risks, it is useful to refer to standard texts or reviews such as, for example, a study group report by the Royal Society on risk assessment.

4.4.3   Presentation of Data

Since it is likely that the reader of a report relating to chemical safety may not be a specialist, it follows that in transmission of results the use of excessive notation and jargon should be avoided. For example, the reader is unlikely to be a statistician, so particular attention should be paid to ensuring that, wherever possible, narrative explanation be provided. It should also be borne in mind that since the reader's mother tongue may not be English, there is a need for the wording used to be as simple and direct as possible and capable of only one interpretation. For example, in reporting results, the conventional significance levels of 0.05, 0.01 and 0.001 are sufficient, and are translated as 'statistically significant', 'highly statistically significant', and 'very highly statistically significant' respectively.

4.4.4   Expression of Results as Tables, Graphs, and Figures

It is usually beneficial to make full use of tabulation and graphical, diagrammatic, and other visually attractive methods, on the basis that 'one picture is worth a thousand words'. Some examples of this are shown in Figures 4.2 to 4.4.

4.5   EVALUATION OF EXPERIMENTAL DATA

The No Observed Adverse Effect Level (NOEL, NOAEL) for a substance under consideration, is the most commonly sought quantitative output of a laboratory

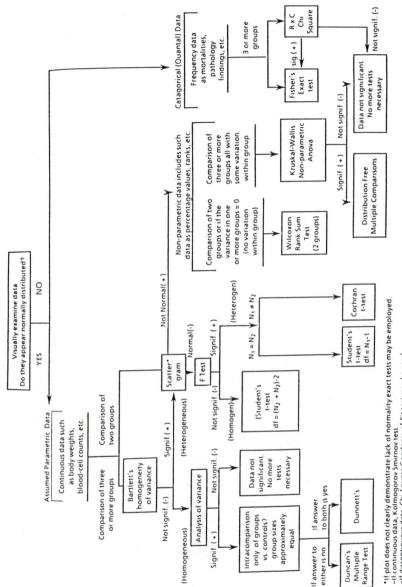

Figure 4.1 *Decision tree for selecting hypothesis-testing procedure.* (*) *If plot does not clearly demonstrate lack of normality, exact tests may be employed.* (+) *If continuous data, Kolmogorov–Smirnov test may be used.* (−) *If discontinuous data, chi-square goodness-of-fit may be used* (Reproduced by kind permission of the publisher from S.C. Gad and C.S. Weil, 'Statistics for toxicologists', in ed. A. Wallace Hayes, 'Principles and Methods of Toxicology'. Raven Press, New York, 2nd edn, 1989)

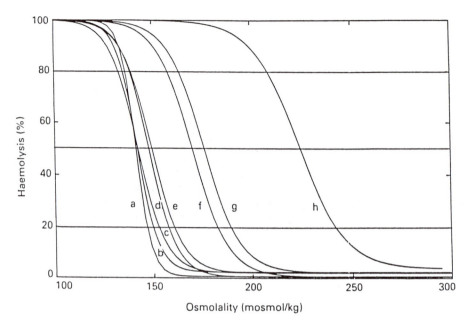

Figure 4.2   *Osmotic fragility curves of erythrocytes from rat, human, and various species of wild artiodactylid mammals: (a)* R. norvegicus, *(b)* H. sapiens, *(c)* E. davidianus, *(d)* C. elaphus, *(e)* B. bonasus, *(f)* C. dama, *(g)* C. axis, *(h)* C. pyrenaica

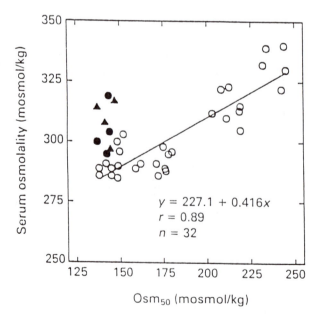

Figure 4.3   *Relationship between serum osmolality and osmolality values that produce haemolysis to 50% of erythrocyte population.* ○ *Artiodactylid species;* ▲ *human;* ● *rat. Regression line for artiodactylid specimens has been plotted and statistical data are included in the Figure*

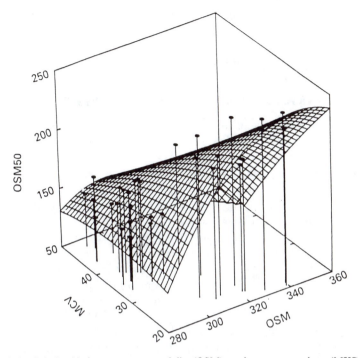

Figure 4.4 *Relationship between serum osmolality (OSM), erythrocyte mean volume (MVC), and Osm$_{50}$ (OSM50) values. The curved surface has been generated by cubic spline fitting. Points represent individual values, and lines under each one are plotted for greater clarity*

experiment, being the 'threshold dose' below which the body is thought to be able safely to dispose of a xenobiotic.

Based on estimates of exposure and knowledge of the NOEL, the estimation-of-risk may progress, for example in respect of a food substance or environmental contaminant, by calculation of an Acceptable Daily Intake (ADI) or Threshold Limit Value (TLV). These are not to be regarded as sacrosanct figures; for example the FAO/WHO (Food and Agriculture Organization/World Health Organisation) Joint Expert Committee on Food Additives states 'An ADI provides a sufficiently large safety margin to ensure that there may be no undue concern about occasionally exceeding it, providing that the average intake over longer periods of time does not exceed it'.

The process of utilizing data from laboratory experiments in order to make estimates of risk in real-life situations is extrapolation and therefore it is to some extent uncertain. The relative imprecision of the extrapolation process is compensated for by the use of factors, whose effect is to make the final estimate highly conservative: a 10-fold factor is utilized to compensate for possible inter-individual variation (for example to protect a particularly sensitive individual) and a further 10-fold factor is employed to compensate for extrapolations across species. Therefore a 100-fold factor (x100) is utilized when extrapolating from a

NOAEL obtained in a laboratory study in rodents if man is the eventual target species. On some occasions, factors greater than x100 have been used, for example when there have been uncertainties in the data that are available or doubts about the purity of the tested substance.

4.6   ERRORS AND FAULTS IN DATA INTERPRETATION

It is easy to see how, after application of such conservative factors, the calculated ADI or TLV are in practice not regarded as immutable numbers. Indeed one of their purposes may be to focus attention and to facilitate forward planning. On this basis the numerical values for allowable exposure may be adjusted (often downwards for environmental contaminants, sometimes upwards for beneficial therapeutic entities) as more precise, relevant data are generated and as overall knowledge regarding the entity increases.

Whilst these adjustments to extrapolated values are properly to be regarded as part of the iterative phase of safety assessment, there are other situations where errors and faults of data handling can jeopardize the data interpretation process. For example when two groups are compared with a view to detecting a significant difference between them (using the null hypothesis) there is never a completely certain answer. Only a probability. The degree of certainty is given by the P value. When the P value is 0.01 then there is only a 1% likelihood that the difference is due to chance variation, which means that, on average, in 100 studies demonstrating a significant difference at this level, there will be one occasion when the difference is not real but is accidental and so would be a falsely positive result.

False positive results arise when test systems are unnecessarily oversensitive. Such results will lead to overestimates of hazard and unduly low NOAELs, which when extrapolated give overestimates of actual risk. Whilst such errors are 'on the safe side', they may prove to be costly in other ways, for example by leading to over-investment on containment (in the case of an industrial chemical) or possibly leading to underdosage during the development of a new pharmaceutical entity.

At least as unsatisfactory an issue as 'false positives' is that of false negative results. This is because an underestimate of potential for hazard can have serious consequences. Probably the most common reason for failure to detect a difference between two groups, even when one exists, is that the experiment is too small. For example, the number of patients who must be enrolled in a trial investigating a new candidate drug will depend on:

–   the degree of variability of the expected data
–   the minimum size of the relevant measurement that will be made
–   the P value sought, which is usually 0.01 or 0.05
–   the power of the study, usually 80% or 90% is required

4.7   SUMMARY

The following are useful questions to ask when interpreting data.

- Are the data valid, that is, do they result from application of approved methodology, competently performed?
- Have the data been validated, that is, have they been subject to quality assurance audit (*e.g.* all modern regulatory toxicological studies); and is there available a peer review (*e.g.* there is in many modern toxicological studies which contain subjective assessments such as pathology)?
- Has the experiment result been proved to be reproducible (*e.g.* performed in duplicate on separate occasions, the standard approach for many of the *in vitro* genotoxicological tests)?
- Has the chosen validation process been adequately documented: for example, have the data been generated under Good Laboratory Practice conditions (GLP) or under a National Measurement Accreditation Service (NAMAS) scheme?
- Do the data arise from a sufficiently sized experiment [modern regulatory toxicological study designs are harmonized on an international basis, according to the Organisation of Economic Communities (OECD) or, for pharmaceutical guidelines, the International Conference on Harmonization (ICH)] .
- Did the experiment yield sufficient controlled data to allow within-study comparisons to be made (assuming adequate study conduct, the accepted designs usually yield sufficient data to enable interpretation of the study without resorting to use of background data)?
- Are the data from the experiment in agreement with relevant background data (although it should be possible to interpret the study data without recourse to other data, the availability of concurrent background data, which has been generated at the same laboratory, can be invaluable as a means of validating the study under review; such data can also be helpful by putting the study results into a proper context)?

If all of the above are satisfactory, then, the final question can be asked:

- Do the data arising from the test group(s) differ quantitatively and/or qualitatively from the control groups and relevant background data? If yes, then this may be attributed to an effect of the test substance.

4.8   BIBLIOGRAPHY

G.W. Bradley, 'Disease, Diagnosis and Decisions'. Wiley, New York, 1993.

D.J. Hand, F. Daly, K. McConway, D. Lunn, and E. Ostrowski, 'A Handbook of Small Data Sets'. Chapman and Hall, London, 1994.

R. Johnson and G. Bhattacharyya, 'Statistics – Principles and Methods'. Wiley, New York, 1992.

D. Moore and G. McCabe, 'Introduction to the Practice of Statistics'. W.H. Freeman, San Francisco, 1993.

'Risk Assessment. A Study Group Report of the Royal Society'. Royal Society, London, 1983.

CHAPTER 5

Risk Assessment

H. PAUL A. ILLING

Risk assessment and risk management are impossible to separate. Risk assessment is essentially a preliminary to setting up proper risk management procedures. Confirmation of the effectiveness of risk management procedures requires reassessment of the risks after implementation of these procedures. Thus, although principles and procedures for risk assessment will be discussed in this chapter and principles of risk management in the next, in practice the two chapters need to be read together.

5.1 DEFINITIONS

Formal definitions of risk, risk assessment, and risk management are contained in the 'Glossary for Chemists of Terms Used in Toxicology' (IUPAC, 1993). Risk is:

1. the possibility that a harmful event (death, injury, or loss) may occur under specific conditions as a consequence of exposure to a chemical or physical agent,
2. an expected frequency or occurrence of a harmful event (death, injury, or loss) arising from exposure to a chemical or physical agent under specific conditions.

Risk assessment is:

the identification and quantification of the risk resulting from the specific use, or occurrence, of a chemical or physical agent. Such assessment must take into account the possible harmful effects, on individuals or society, of using the chemical or physical agent in the amount and manner proposed considering all possible routes of exposure. Quantification ideally requires the establishment of dose–effect and dose–response relationships in likely target individuals and populations.

Risk management is:

The decision making process, taking into account political, social, economic, and engineering factors as well as the potential hazard, thus developing, analysing, and comparing regulatory options enabling the selection of the optimal regulatory response for safety from that hazard. Essentially risk management is the combination of three steps: risk evaluation, emission and exposure control, and risk monitoring.

Although these are definitions recommended by IUPAC, many other definitions are in widespread use. For example, risk is 'the possibility of suffering harm from a hazard' (Cohrssen and Covello, 1989). This introduces two further concepts, the harm associated with the hazardous substance or process, and the likely exposure to that hazardous substance or process. The IUPAC definitions of hazard and exposure are:
Hazard:

The set of inherent properties of a substance, a mixture of substances, or a process involving substances, that, under production, usage, or disposal conditions, make it capable of causing adverse effect to organisms or the environment (*i.e.* harm), depending on the degree of exposure. In other words it is a source of danger.

Exposure:

1. The concentration, amount of intensity of a particular physical or chemical agent or environmental agent that reaches a target population, organism, organ, tissue, or cell. This is usually expressed in numerical terms of substance, concentration, duration, and frequency (for chemical agents and micro-organisms) or intensity (for physical agents such as radiation).
2. The process by which a substance becomes available for absorption by a target population, organisms, organ, tissue, or cell, by any route.

5.2   PROCESS OF RISK ASSESSMENT

The process of risk assessment can be divided into four stages, as follows.

5.2.1   Hazard Identification

Determination of substances giving rise to concern, and their adverse effects.

5.2.2   Dose–Effect or Dose–Response Assessment

Determination of the relationship between dose and severity of effect (dose–effect) or dose and frequency of effect (dose–response) (see below).

5.2.3 Exposure Assessment

Measurement or estimation of the intensity, frequency, and duration of exposure to an agent. (This may require information on source and magnitude of potential release(s) and on dispersion patterns, or it may be measured or modelled directly.)

5.2.4 Risk Estimation

The process of combining the information, in order to quantify the risks to a population.

Some harmful effects, notably cancers, are considered to be 'stochastic', either they are present or they are not. Most effects, however, are 'non-stochastic' and the severity in an individual increases progressively with dose. Non-stochastic health effects commonly occur through the following process:

Normal → Homeostatic adjustment → Compensation (input of reserve capacity with possibility of repair) → Breakdown (leading to increased disability and ultimately death).

A similar progression can be described for environmental effects, although the most severe stages will be 'disorganization' and 'disintegration' rather than disability and death.

If the inter-individual variation of the severity of the effect with dose is small, then it is possible to determine a dose–effect curve for the whole population. However, if the variation is large, population comparisons are based on the dose required to cause a fixed level of effect in each individual in the population, and are therefore assessments of dose–response. The relationship between these steps can be shown diagrammatically in Figure 5.1.

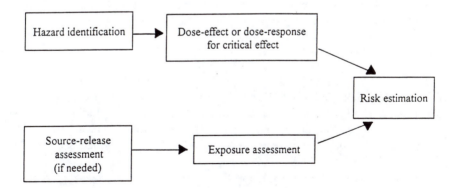

Figure 5.1 *Stages in carrying out risk assessment*

5.3   IDENTIFICATION OF HAZARD

Identifying the potential harm of a substance or process may involve all or any of the following: observation, experimental work, information retrieval, and deductive work based on physicochemical parameters and structure–activity relationships. In the past, much of this information was retrieved from peer-reviewed articles in recognized journals. With the advent of internationally agreed standard test protocols and the auditing requirements of Good Laboratory Practice (GLP), a study conducted using these protocols, and subjected to GLP quality assurance procedures, should be considered as being of acceptable standard. However, such studies are often difficult to retrieve. They may only be available from the sponsor of the test, in which case they will not be entered into the databases used for literature searches.

Hazard identification and determination of dose–response or assessment is relatively easy for physical effects such as fire or explosion, but much more difficult for toxicological effects. Data on effects on human health may come from the following sources:

–   Human observations (case reports, epidemiological studies, or, in suitable cases, experimental human studies)
–   Animal toxicological studies
–   Assessment of structure–activity relationships

The study data can be supplemented with toxicokinetic and toxicodynamic information to improve the interpretation of data and to confirm (or otherwise) the assumption (usually made), that the behaviour in the test animal species is similar to that in the target species.

Evidence of harm to the environment may come from:

–   assessment of structure–activity relationships
–   toxicity studies in indicator species
–   studies using microcosms or mesocosms
–   field studies and field observations

Predictive evidence for environmental effects is usually obtained in stages, with the more complex studies only being undertaken if there is a specific need. The information available on chemical hazards from 'pivotal' studies (*i.e.* those identifying the 'critical' effect) may be incomplete or difficult to interpret. The pivotal study may depend on the exposure situation being examined (see below).

5.4   EXPOSURE ASSESSMENT

There are two main circumstances where exposure information is required, accidents and occasions where low-level exposure is expected (anticipatable exposure). Many situations are possible, depending on intended methods of synthesis, intended uses, likely disposal routes, *etc.* Likely exposure levels will depend on a number of factors. Possible sources of exposure are given in Figure 5.2.

Figure 5.2 *Potential exposure scenarios for consideration in risk assessment*

A model has to be established with which to predict likely exposures following a major accident. The important factors affecting this exposure model are the source term and the dispersion patterns. The frequency part of the source term can be obtained by 'fault tree' analysis, identifying the consequences of failure *etc.*, or 'event tree analysis', whereby potential events are analysed to identify what failures could cause them. Alternatively, engineering judgement based on historical event frequencies allow employees to estimate appropriate failure rates. The size and duration of the postulated or actual release resulting from these failures is then calculated and combined with the failure information to give the overall source term. The dispersion of the toxic cloud released is modelled using a knowledge of the buoyancy of the cloud and weather conditions, in order to obtain a series of concentration–time relationships known as 'isopleths'. At a given time point this concentration–time combination will exceed that expected. An overall isopleth envelope can then be calculated for a given set of release conditions by combining the individual isopleths for each time point post-event. This is illustrated diagrammatically in Figure 5.3.

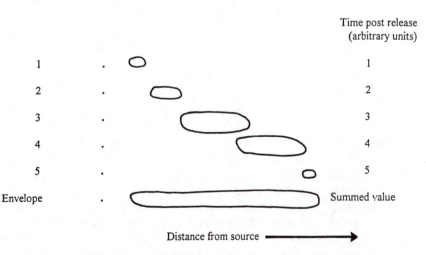

Figure 5.3 *Development of exposure isopleths for dispersion following a point release of a chemical. The isopleths for each individual time post-release are combined to give an overall isopleth for the release*

A risk contour may be required for the probability of exposure to a given concentration–time combination of the substance. In this case the isopleth envelopes are combined with information on weather patterns in order to determine the overall area in which there is a defined probability of the relevant isopleth envelope occurring during a given time period.

For anticipated exposure, direct measurement or modelling of exposure is used. It may be necessary to examine several situations in order to evaluate the risks throughout a life cycle ('Life cycle analysis'). Exposure may occur during synthesis, use in manufacturing, use as a product, or disposal. An example of such a situation is that of a drug, where the workforce may be exposed during synthesis and formulation, the patients and hospital staff during use, and the general population during storage and disposal, either of the drug or its metabolites (*e.g.* in patients' excreta). Food basket studies can be used to obtain a measure of how much of a chemical is likely to be ingested, if present in foods.

For some chemicals, the measurement of levels in the workplace, in foods or in a particular environmental compartment (water, air, soil, *etc.*) may be necessary. In many circumstances modelling exposure to chemicals will be all that is initially possible. However, if as a result of modelling it seems that exposure is likely to be in excess of acceptable levels, then further measurements may be necessary.

5.5   RISK ESTIMATION

5.5.1   Major Accident Exposure

The risk defined in the section on exposure is one of being exposed to a given concentration–time combination (*i.e.* a parameter of exposure). This becomes a harmful risk when the concentration–time relationship indicates that a toxic effect is likely to occur (*i.e.* represents individual risk, even though population-based data are used in deriving it). If a risk decision is to be made on the size of the potential affected population (*i.e.* a societal risk), then the population density in the affected area will also need to be included in the considerations. Depending on the harm criterion used, the risk estimate obtained may be employed either for decisions on land use or for emergency planning purposes.

Toxicity data may be transformed into a concentration–time–effect relationship, which is suitable for describing the level of harm for which predictive information is required. For this sort of predictive work, the best estimates are required, thus conventional 'safety' (uncertainty) factors are usually omitted. The estimates are usually for 'boundary conditions', *e.g.* for conditions where only a very small proportion of the affected population will die, rather than for mean values such as the LCT_{50}. This is the lethal concentration–time combination which results in 50% mortality and is analogous to the median lethal concentration (LC_{50}) (*see* Appendix 2). Generally, animal data have to be used but, where available, human data should be used to confirm the validity of the animal data. If several species have been studied, the most sensitive is chosen to represent man unless this is considered inappropriate. Extrapolation of data from the mean value to the relevant boundary value may need to be undertaken.

The data on harm are linked to data on exposure to yield the risk estimation. Finally, an analysis of the overall uncertainty should be undertaken in order to indicate how robust the estimate is.

5.5.2 Other Exposures: Human Health

Usually, the risk consequences arising from exposure during minor misuse and small-scale accidents, and the consequences arising from normal use, are considered when conducting this type of risk estimation. Generally, the estimation is carried out by obtaining the exposure level derived as an acceptable ('safe') level, based on examination of data on harm, and comparing with the measured or modelled exposure data.

A severe effect should occur only very rarely, whereas a mild effect can be allowed to occur more frequently. It is also essential to take into account the type of population studied (*e.g.* inbred, young healthy male rats) and the type of population at risk [*e.g.* a working population (male, female, heterogeneous, mainly healthy, adult) or a general population (all ages and sexes, including the young, the elderly, and the infirm)]. That is, it is necessary to take into consideration inter-species differences and inter-individual differences.

5.5.3 Other Exposures: Environmental Effects

The principal methods used predictively are based on identifying the most important effect in the relevant environment. The approach depends on comparing the Predicted No Effect Concentration with the Predicted Exposure Concentration and ensuring that there is a sufficient margin between these two values. Confirmation of the validity of the estimates may be obtained from field observation or study. This is usually conducted for a specific reason, for example to ensure that serious disturbance is not caused to non-target 'beneficial' organisms when a chemical is used as a pesticide on field crops.

5.6 EXTRAPOLATION FROM TOXICITY DATA TO ACCEPTABLE EXPOSURE LEVELS

5.6.1 Non-stochastic effects

For non-stochastic effects, extrapolation is usually made using the No Observed Adverse Effect Level (NOAEL) for the critical effect. In doing this an assumption is made but never explicitly stated. It is that the combination of an acceptably low frequency and severity of harm reduces an average dose–effect curve to a constant. The International Union of Pure and Applied Chemistry (IUPAC) recommends the World Health Organisation definition of a NOAEL:

> The greatest concentration or amount of a substance, found by experiment or observation, which causes no detectable adverse alteration of morphology,

functional capacity, growth, development, or life span of the target organism under defined conditions of exposure.

The NOAEL is adjusted using 'safety' (uncertainty) factors in order to determine the 'Health Based Occupational Exposure Limit' (for workplace exposure), a scientifically based 'Environmental Quality Standard' (for environmental exposure), an 'Acceptable Daily Intake' for food additives, or a 'Tolerable Daily Intake' for food contaminants. Conventionally, extrapolation from a no-effect level in a well-conducted study in rats to a human general population is based on a factor of 100 (10 for species, 10 for individual variation). Expert judgement by specialists in the relevant subject areas is then used to modify that factor up or down. Toxicokinetic and toxicodynamic studies are often used to assist in that judgement. These factors may be enlarged to deal with NOAELs based only on absence of severe effect. Very small factors may be applied when extrapolating from one human population to another for workplace exposure and minor effects such as sensory irritation.

5.6.2   Stochastic effects

The principal stochastic effect is that of cancers. Risk assessment for stochastic effects is based on frequencies. Many systems used to classify carcinogens depend on quality of evidence, not on potency (although more potent carcinogens are more likely to be identified), and are therefore not categorization systems for risk estimation. The ill effect in cancers is severe, and extrapolation to very low frequencies is required. In the USA, 'quantified risk assessment' is attempted. This may be carried out on data from studies in humans or animals. Several mathematical models are used and the models and the data used are clearly identified. In the UK it is felt that a number of questionable assumptions are often made (Committee on Carcinogenicity of Chemicals in Food, Consumer Products and the Environment, 1991) namely:

- they are derived from mathematical assumptions rather than from a knowledge of the biochemical mechanisms;
- they often use incomplete or inappropriate data;
- they currently demonstrate a disturbingly wide variation in risk estimates, depending on the model used, and are therefore not validated.

In these circumstances it is the view of UK experts that judgement must be relied on when making extrapolations of frequencies to levels of exposure which are predicted to give very low frequencies of occurrence of effect.

5.7   CONCLUSIONS

In this chapter the technical aspects of risk assessment have been considered. It has been demonstrated that it is impossible to consider risk estimation without considering risk evaluation, i.e. the setting of risk estimates against criteria.

Although conceptually different, these cannot be separated in practice. Development of suitable criteria for use in the risk evaluation depends on societal decisions on what is an 'acceptable' or 'tolerable' risk. Thus the process of managing risk moves away from a purely technical exercise to one that also takes into account public perceptions. The framework in which the criteria and the decision taking are set, and the management procedures which flow from the risk evaluation, are discussed in the next chapter.

5.8 BIBLIOGRAPHY

B. Ballantyne, T. Marrs, and P. Turner, 'General and Applied Toxicology. Macmillan, London, 1993.

J.J. Cohrssen and V. Covello, 'Risk Analysis: A Guide to the Principles and Methods for Analysing Health and Environmental Risks, National Technical Information Service, Springfield, VA, 1989.

Committee on Carcinogenicity of Chemicals in Food, Consumer Products and the Environment, 'Guidelines for the Evaluations of Chemicals for Carcinogenicity', Department of Health Reports Health Social Subjects 42, London, 1991.

Committee on Risk Assessment of Hazardous Air Pollutants, 'Science and Judgement in Risk Assessment', National Academy Press, Washington, DC, 1994.

S. Fairhurst, The uncertainty factor in setting occupational exposure standards, *Ann. Occup. Hyg.*, 1995, **39**, 375–384.

IUPAC, Glossary for chemists of terms used in toxicology, *Pure Appl. Chem.*, 1993, **65**, 2003–2122.

V.C. Marshal, 'Major Chemical Hazards'. Ellis Horwood, Chichester, 1987.

A.G. Renwick, Data-derived safety factors for the evaluation of food additives and environmental contaminents, *Food Addit. Contam.*, 1993, **10**, 275–305.

A.G. Renwick, The use of an additional safety or uncertainty factor for nature of toxicity in the estimation of Acceptable Daily Intake or Tolerable Daily Intake. *Regulatory Toxicol. Pharmacol.* 1995, **22**, 250–261.

M.R. Richardson, 'Chemical Safety'. VCH, Weinheim, 1994.

CHAPTER 6

Risk Management

H. PAUL A. ILLING

The previous chapter was concerned with the technical and scientific process of risk assessment. This one is concerned with the stages of risk evaluation and management. These processes require societal input, both in terms of the perceptual and legal frameworks in which decisions are taken and in terms of the criteria that lie behind the decisions taken. Thus, this chapter is concerned with matters which transcend scientific and technical knowledge.

6.1   RISK CONSIDERATIONS

The basis for a regulatory process involving a control strategy taking into account cost–benefit considerations was set in the UK in the report of a study group set up by the Royal Society in 1983. This group established:

- an upper limit of risk that should not be exceeded for any individual.
- further control, so far as is reasonably practicable, making allowance if possible, for aversions to the higher levels of risk detriment.
- a cut-off in the deployment of resources below some level of exposure or detriment judged to be trivial.

These parameters can be set out diagrammatically as shown in Table 6.1 below. When dealing with quantified risk assessment, these definitions need to be linked to some form of numerical value for a defined effect. However, many risk assessments are qualitative, and often the criteria used are not explicitly stated. It

Table 6.1 *Outline of Royal Society approach to risk management*

Broadly acceptable	Tolerable	Unacceptable
	Willingness to live with a risk so as to secure certain benefits and with confidence that it is being properly controlled.	

INCREASING RISK ➤

62

is also necessary to distinguish between individual and societal (or population) risk when drawing conclusions.

Individual risk is independent of population density, while the societal (or population) risk depends on the population at risk and therefore considers the population contained in a geographical area. Individual and societal risks are important when considering the possibility of accidental exposure. Individual risk is the usual form for dealing with small-scale accidental and anticipatable risk. Population risk is usually the more important form of risk for environmental effects.

6.2 CRITERIA FOR RISK EVALUATION: MAJOR ACCIDENTAL RISKS AND HUMAN HEALTH

Although comparisons of the risks from different sources are based on premature death due to an agent, this is usually too severe for risk management purposes. A range of criteria is used for major accident hazards.

6.2.1 Emergency Planning

One set of definitions of criteria levels is implicit in a categorization scheme developed by the European Centre for Ecotoxicity and Toxicology of Chemicals (ECETOC) in 1991 for emergency planning. Four zones are defined, based on different categories of risk.

- Death or permanent incapacity. This is the most severe category, death or permanent incapacity occurring immediately or shortly after exposure, and includes severe effects, such as permanent (unless surgically corrected) blindness.
- Disability. This includes individuals who are markedly helped by external assistance, and treatment results in full recovery.
- Discomfort. The discomfort category includes those for whom a full recovery is probably without external assistance, although systematic relief may be possible and reassurance desirable.
- Detectability. This is a category where sensory irritation may occur (*e.g.* a substance detected by its unpleasant smell), but there is no direct effect of exposure on health.

These are essentially definitions of the categories, not definitions of the boundary criteria between the categories, but boundary criteria are implicit in the definitions.

6.2.2 Land Use Planning

For land use planning a single category, and therefore criterion, is required. This is usually a severe criterion, similar to that used for the death or permanent incapacity category of ECETOC. One definition established by the Health and Safety Executive in 1989 requires:

– Severe distress to almost everyone
– A substantial fraction requiring medical attention
– Some people seriously injured requiring prolonged treatment

or

– Highly succeptible people might be killed

In this case the defined risk criterion is the boundary criterion.

6.3   CRITERIA FOR RISK EVALUATION: ANTICIPATABLE RISK TO HUMAN HEALTH

The criteria for anticipatable risks are essentially aimed at achieving very low levels of risk. Health-based or scientifically based approaches are associated with a combination of frequency and severity which is considered broadly acceptable. Often a figure of excess deaths of 1 in a million (lifetime risk) is quoted for cancers (*i.e.* a stochastic risk of a severe effect) when undertaking model-based risk extrapolations. Excess death is defined as that due to anticipatable risk in excess of an incidence due to other causes. Those who do not accept model-based approaches usually suggest that this level of risk cannot be determined. Broadly acceptable frequencies for different levels of milder ill-health associated with conventional, non-stochastic, toxic effects have not been stated explicitly. They are taken into account (implicitly) when considering the safety factors applied to the No Observed Adverse Effect Level (NOAEL). This approach leads to, for example, the Acceptable Daily Intake or the UK Occupational Exposure Standard.

6.4   CRITERIA FOR RISK EVALUATION: ENVIRONMENTAL RISK

Criteria for environmental risk are still poorly defined. There is clearly a gradation of risk according to the environment being affected. Effects on areas of particular environmental importance are minimized to a greater extent. There also has to be a trade-off between different risks (environmentally mediated human health effects and effects on other species/ecosystems). How to deal with the balance between these risks is a societal judgement, and consensus concerning that judgement is not available. It may vary according to the specific stiuation.

6.5   TOLERABLE RISK

The concept of tolerable risk carries with it a view that there is a benefit associated with, and outweighing, the risk. Decisions concerning these risks are not solely based on scientific evaluation; they also depend on the practicality of possible risk management procedures. Phrases such as As Low As is Reasonably Achievable (Practicable) [ALARA(P)], Best Available Technique Not Entailing Excessive Cost (BATNEEC) and Best Practicable Environmental Option (BPEO) are intended to cover this requirement. This can lead to the setting of a minimum standard considered achievable by everyone, somewhere within the band of tolerable risk, and with a continuing duty to improve. The UK Maximum Exposure Limit for occupational exposures is one such limit.

6.6 CRITERIA FOR RISK EVALUATION: FURTHER COMMENTS

The view that a risk is broadly acceptable, tolerable, or unacceptable, and the decisions concerning appropriate risk management, are essentially societal decisions. They depend on public perception and opinion. Consequently, individuals and groups of individuals within that society may not agree with the decision arrived at and may seek to change public perception.

The approach outlined above presumes that the evidence available is complete and convincing. Often this is not the case, and where evidence is incomplete it may not be reasonable to try to obtain it. In such cases, it may be necessary to adopt the precautionary principle and to incorporate additional factors to allow for the uncertainties of the information. This may take the form of adopting a lower limit of exposure as broadly acceptable, or to treat a risk as unacceptable. If further evidence can be gathered, then the reduction in uncertainty concerning the risks could properly lead to a re-evaluation of how they should be managed.

6.7 THE RISK EVALUATION PROCESS

At this point it is necessary to examine concepts behind potential governmental approaches to risk evaluation and management. Specific legislation of individual governments will not be dealt with here, only the general principles.

The starting point for consideration is national governments. These can elect to take decisions directly or to set up means of delegating the decisions, either upwards to international bodies such as various United Nations organizations, or downwards to local government, companies, or individuals. International companies can act across national boundaries by imposing their own higher standards on local subsidiaries, but these have to be set in a manner acceptable to the countries in which they are operating.

Legislation is passed by governments who set up a decision making structure and a general framework for risk management. The legislation may establish an organization that considers the detailed technicalities of the problem and comes to a decision on whether to allow certain risks to be taken, and what detailed risk management procedures should be undertaken. Alternatively, legislation may leave the legal liability associated with the consequences arising from the use of certain chemicals with the producer, or user, or consumer. In these circumstances, the producer or user performs the risk evaluation. Generally, legislation tends to follow governmental department boundaries. In scientific and technical terms these boundaries are artificial, and the consequence is that the use of a chemical may have to comply with more than one set of legislation in the same country.

Legislation may cover safety in the workplace, safety of the product, or environmental protection. As these areas can overlap it is essential to ensure that legislation does not result in mutually opposed management requirements. Indeed, environmental safety often has to be considered in conjunction with either workplace safety or product safety (or both) as illustrated in Figure 6.1.

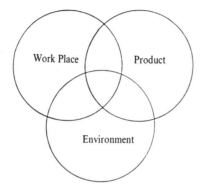

Figure 6.1   *Interaction of legislation for safety in the workplace, in the environment, and of the product*

Decisions on risk evaluation and management may be taken by consensus or confrontationally. In a consensus-based procedure, technical aspects of risk evaluation are often left to expert committees, consisting of people appointed for their technical knowledge and experience, or, for less contentious decisions, to single experts. The wider societal input to decision taking can be made by consensus through representatives of the relevant interest groups (*e.g.* trade unions and industry leaders in workplace safety, *etc.*) acting on behalf of society as a whole.

In a confrontational approach, all views on a problem may be presented using a judicial or quasi-judicial process, and an appointed judge (or arbitrator) assesses the views expressed and either delivers an opinion to be promulgated by the appointing authority or, if the authority has been fully delegated, the decision. The latter approach is frequently used for decisions concerning the requirements of the need for a particular chemical process which may constitute a major hazard. The judge (arbitrator) would consider such problems as where to site the plant, what further developments to allow in the vicinity of the plant, *etc.*

6.8   RISK MANAGEMENT

Risk management is concerned with the consequences of risk evaluation. When dealing with broadly acceptable risks or unacceptable risks, no further management action is required, except where there is a ban which needs to be enforced. However, decisions may be much harder when dealing with tolerable risk. Ideally, management is required to reduce tolerable risks to the point where the risk is broadly acceptable. This may not be possible, but at least it should be reduced as far as is reasonably practicable.

The approaches used for safety in the workplace differ fundamentally according to the type of risk (*i.e.* for risks arising from major accident hazards or for risks from minor accidental and anticipatable exposure) when dealing with tolerable risks.

6.8.1 Major Accident Hazards

For major accident hazards the risk management consists of two parts. The first is planning by ensuring that risks are minimized through appropriate choice of location. This is land use planning (see Section 6.2.2). There are two elements in land use planning: where to site the plant and what to do about permitting developments of factories, housing, *etc.* around the plant. The second consideration is the requirement for safety to be taken into account in designing the process (the chemical process). Thus, legislation may be used to ensure that 'safety cases' covering the process design and the plans for emergency procedures, including environmental cleanup, are taken into account.

6.8.2 Anticipatable and Minor Accident Workplace Exposure

For minor accidents or anticipated exposure, the approach to controlling risks in the workplace may be summarized as follows:

- Substitution: use another, less risky process or chemical.
- Engineering control: change the process to prevent/reduce exposure, use ventilation if process design improvement is inappropriate.
- Use of personal protective equipment: hard hats, eye protectors, clothes, gloves, footwear, *etc.* including respiratory protective equipment (*e.g.* face masks, airstream helmets).

As a solution to a problem, substitution is preferred to engineering control, which is preferred to protective equipment.

These control measures are used in combination with exposure limits. Exposure above a limit based on tolerable risk is normally not permitted. If a limit set on the basis of broadly acceptable risk is exceeded, then ways are sought to reduce exposure below that limit.

Proper maintenance of processes and equipment is essential to ensure that they function correctly, and occupational hygiene measurements and health surveillance may be needed to ensure that the controls are adequate. These measures are only fully effective when set within a suitable management system in an organization that is safety oriented.

6.8.3   Controlling Effects on the Environment

Management of environmental risk is concerned with prevention and 'cleanup'. Ideally, prevention is preferred. Once pollution has taken place, restoration of the environment to its former condition is desirable. The 'polluter pays' principle defines one approach to providing the means to execute the restoration.

6.8.4   Product Safety

Product safety is concerned with whether the use to which the product will be put is safe. One way of achieving this is to pass on hazard information in order to enable the recipient to carry out a risk assessment. Another is for the organization marketing the product to carry out its own risk assessment, based on the likely behaviour of the recipient. If society considers that such a matter is sufficiently important, then a decision may be taken by government or even internationally, that will set a broadly acceptable risk level. Where there is a consensus on what constitutes a broadly acceptable risk, the use of internationally derived values can be useful. For example, the Joint Evaluation Committee for Food Additives, a working group of the United Nations Food and Agriculture Organization, collects and correlates such data.

There are circumstances where tolerability is a relevant consideration when dealing with product safety. This is usually where some form of risk–benefit evaluation is undertaken. Two examples illustrate this. Firstly, government may permit the use of a drug with a high incidence of side effects on seriously ill patients. This could be done, after careful judgement by the specialist charged with the patient's care, where the overall benefit to the individual is likely to outweigh the risk. Secondly, a food or drinking water additive may be intentionally introduced in order to prevent undesirable effects such as food spoilage or dental decay. The latter example is concerned with a societal judgement concerning individual risks. It can result in controversy if a significant number of individuals perceive the risks differently from that of the decision taking body.

6.8.5   Enforcement

Enforcement of the outcome of risk management decisions may be through disciplinary committees of relevant professions or groups, or by statutory or insurance-based inspection. The latter may lead to sanction taking such as fines following a court appearance or increased insurance premiums. Where failures do occur, the incidents should be investigated, lessons learned, and, if appropriate, retribution sought. Statutory bodies may instigate this process, or it may be the consequence of individuals seeking redress through a judicial process.

6.9   CONCLUSIONS

Risk management is a difficult area, as it involves a mixture of scientific, technical, and legal processes, and needs to allow for differences between the scientific and technical perception of risk, and the public perception. However, in the final analysis, the public attitude is paramount. There is a clear need to communicate to the general public the value and the limitations of the scientific and technical appraisal in order to ensure that the decisions are taken with full knowledge of all the facts that are available. Communication in relation to risk is in its infancy, but it is essential if the risks associated with chemicals are to be properly understood.

6.10 BIBLIOGRAPHY

Dept. of Environment, 'A Guide to Risk Assessment and Risk Management for Environmental Protection'. HMSO, London, 1995.

'Emergency exposure indices for industrial chemicals', Technical Report No. 43. European Centre for Ecotoxicity and Toxicology of Chemicals, Brussels, 1991.

Health and Safety Executive, 'Risk Criteria for Land Use Planning in the Vicinity of Major Industrial Hazards'. HMSO, London, 1989.

Health and Safety Executive, 'The Tolerability of Risk from Nuclear Power Stations'. HMSO, London, 2nd edn., 1992.

'Risk Assessment. A Study Group Report of the Royal Society'. Royal Society, London, 1983.

CHAPTER 7

Exposure and Monitoring

DOUG TEMPLETON

7.1 INTRODUCTION

In occupational medicine, exposure and monitoring refer to the various means of assessing exposure to harmful or potentially harmful substances in the workplace, their intake and accumulation, and the health risks entailed. Exposure in the environment of the workplace is referred to as ambient exposure. Ambient exposure to a substance that is absorbed—generally through the lungs or mucous membranes of the upper airways, across the skin, or in the digestive system—gives rise to an internal dose that can be defined as the amount of a substance taken into the body. The ultimate goal of monitoring is only realized when a marker is identified that can be measured accurately and which links exposure to the substance with predictable adverse effects at a given internal dose. Medical assessment of the adverse effects themselves (*e.g.* the development of lung disease or malignancy) is beyond the scope of monitoring and is part of a programme of health surveillance. However, when there is the potential for harmful exposures, monitoring programmes and health surveillance go hand in hand.

The distinction between health surveillance and monitoring may be blurred, and the monitoring programme may consist in part of additional investigations that are not part of a routine annual medical examination, which may or may not give a positive result because of exposure to the substance of concern. For example, workers in the hard metal industries may under some circumstances have an annual medical examination that includes evaluation of selected pulmonary function tests and blood samples taken for serum cobalt measurement. A small number of these workers will develop interstitial pulmonary disease but the link, if any, with cobalt is not well understood. In this example, the pulmonary function tests can be considered part of health surveillance in relation to all aspects of employment, while cobalt measurement is part of a monitoring programme specifically for cobalt exposure.

This chapter considers some basic definitions and aspects of monitoring, the criteria for development of a successful monitoring programme, the expanding role of biomarkers in monitoring, and some ethical considerations that arise.

Examples of monitoring exposures to organic and inorganic substances are used to illustrate these principles.

7.2   GENERAL PRINCIPLES

It is usual to distinguish between ambient monitoring and biological monitoring of exposure. In the former, the workplace is monitored, for example by air sampling for the presence of dusts, swipe tests for radiation contamination, *etc.* From the results, inferences about exposure of the whole work force may be drawn. Biological monitoring of exposure, however, is intended to assess the internal dose of the individual worker. In some cases ambient monitoring gives a good indication of individual exposure, while in others even personal samples do not give a very good indication of the absorbed dose. For example, several independent studies from different countries conclude that exposure to soluble compounds of cobalt at 0.05 mg m^{-3} in the workplace air consistently produces a concentration of cobalt in workers' urine of about 30 μg l^{-1} at the end of a shift. The major route of exposure in these studies is inhalation. In contrast, exposure to cadmium results in quite different internal doses in individual workers. Here, ingestion is an important additional route of exposure and individual differences in hygiene (*e.g.* washing the hands before eating) and hand-to-mouth activity (including handling cigarettes) determine quite different internal exposure levels even when personal sampling devices indicate similar ambient exposures.

A third stage of monitoring is monitoring of effects. Here, biological effects are sought that arise from exposure but that are themselves reversible and (or) without serious consequences for health. It must be stressed that this is quite distinct from diagnosis of occupational diseases. The latter represent a failure of preventive programmes, a major component of which may be monitoring. However, as noted above, this distinction, although clear in principle, is not always so in practice. Effects thought harmless at one time may later be found detrimental. New toxic effects are always being elucidated; yesterday's inconsequential exposures may be today's health problems. Individual variation in sensitivity is another reason to be cautious in interpreting effects. The very concept of 'biological effects of no consequence' may be controversial.

Temporal issues as they relate both to exposures and effects are also important in monitoring. Exposures may be short-term or long-term and monitoring has a role in both. Short-term exposures during a work week, a single shift, or even for a very brief time during a shift may all have special consequences. At the other extreme, cumulative exposure over a lifetime may be more significant. High short-term exposures, when they give rise to adverse effects, tend to produce acute toxicity. Cumulative exposure in the long term is more often associated with chronic disease. This is by no means always true. A single brief exposure to a high dose of radiation may produce a chronic leukemia with onset many years later, and exposure to cadmium may present acutely as sudden renal failure after many years of accumulation.

However, closer correlations between exposure and consequence are more usual. This principle is well illustrated by exposures to different compounds of nickel. At one extreme, nickel carbonyl is exceptionally toxic; inhalation of this volatile form of nickel at certain concentrations in air can be fatal rapidly and in cases of accidental exposures is monitored hour-by-hour as urinary nickel. However, survival of acute nickel carbonyl poisoning is without known long-term effects and subtle on-going exposures are apparently without consequence. At the other extreme, inhalation of insoluble nickel compounds in dusts has no immediate consequences and is very difficult to detect by biological monitoring. It does, however, result in deposition of nickel in the upper airways that may be detected on nasal mucosal biopsy even years after retirement from nickel-related occupations, and poses an increased risk of eventually developing cancer in the upper airways. The majority of workers exposed occupationally to nickel fall between these extremes. They are exposed to a mixture of soluble and insoluble forms of the metal, are monitored for during-shift exposure by urinary nickel measurement, and may be (though unproven) at slightly increased risk of developing dermatitis and (or) upper airway cancer.

Several terms are used to describe ambient and internal exposures; among the most commonly used for setting permissible exposure levels are those of the American Conference of Governmental Industrial Hygienists (ACGIH) and the Deutsche Forschungsgemeinschaft (DFG). ACGIH publishes threshold limit values (TLVs) for a number of substances that are guidelines to upper ambient exposures. Because a particular level of exposure may be tolerable in the short term but not for on-going exposure, three different TLVs are used. A time-weighted average (TWA) TLV is the time-weighted average exposure during an 8 h day and 40 h working week to which nearly all workers may be exposed indefinitely without adverse effects. A short-term exposure limit (STEL) is a 15 min time-weighted average exposure which should not be exceeded at any time even if the 8 h exposure is within the TLVTWA. Exposures at the STEL should not be for more than 15 min, not more than four times per day, and not more than once in a 60 min period. A ceiling value (TLV-C) is one which should never be exceeded for any length of time.

DFG uses a maximum concentration value in the workplace (MAK) which it defines as 'the maximum permissible concentration of a chemical compound present in the air which, according to current knowledge, does not impair the health of the employee or cause undue annoyance'. For some substances there is concern that no level of exposure is free from adverse health consequences. For example, if a substance is carcinogenic, it may have no threshold value for increased risk yet exposure in the workplace may be unavoidable. Under these circumstances, DFG uses a technical exposure limit (TRK) based on attainability with current technologies to serve as a guideline for necessary protective procedures. Returning to the above example of nickel, considered by DFG as carcinogenic and therefore presumed by them to be without any risk-free level of exposure, the inevitable use of the metal in industrialized society dictates a realistic approach. DFG believes that current technology can reasonably assure exposures below 0.5 mg m^{-3} while maintaining relevant industries, and sets this as the TRK.

MAK, TRK, and TLVs describe ambient exposures. Internal exposure is addressed by ACGIH with a biological exposure index (BEI) which represents the warning level of a particular indicator in an appropriate biological sample. The comparable DFG term is the biological tolerance value (BAT) which is defined as 'the maximum permissible quantity of a chemical compound [or] its metabolites, or [the magnitude of] any deviation from the norm of biological parameters induced by these substances in exposed humans'. Examples of these different quantities should make this definition clearer. The BAT for aluminum is simply reported as 200 µg l^{-1} of urine; the chemical itself is measured. Carbon disulfide is extensively metabolized and one of its metabolites 2-thiothiazolidine-4-carboxylic acid, (TCCA) is measured in urine. The BAT for carbon disulfide is accordingly given as 8 mg l^{-1} of TCCA in urine. In the case of acetylcholine esterase inhibitors, neither the chemical nor its metabolites are measured. Rather, the BAT is reported as a reduction in erythrocyte acetylcholine esterase activity to 70 % of the reference value. (In each of these examples, the sample is taken at the end of the working shift.)

There is a basic difference between BAT and BEI values, and neither is without problems. The BAT is health-based and assumes sufficient information to decide on levels with no adverse effects. Such information is frequently unavailable. The BEI, however, is the level of the parameter expected in an average worker exposed at the TLV, and so assumes inhalational exposure and absorption. Individual variation means that some workers will have levels above the BEI, and the importance of these internal doses is generally poorly understood.

Other jurisdictions may use their own terminology. Many define Biological Action Levels for substances that conceptually approximate the BEI and BAT.

7.3   CRITERIA FOR A MONITORING PROGRAMME

In order to monitor exposure in a specific instance, or to assess the worth of a monitoring programme, a number of criteria should be examined.

7.3.1   Is Biological Monitoring a Useful Supplement to Ambient Monitoring? What are the Goals of the Monitoring Programme?

If ambient monitoring shows acceptably low levels of contamination, there may be little to add by biological monitoring. However, in some circumstances internal doses may be unacceptable in certain individuals due to additional or unpredictable routes of exposure (*e.g.* oral) and this will only be detected by individual biological monitoring. The goals of the programme need to be defined. Is biological monitoring to be used as an adjunct to ambient monitoring only to assess exposure, or is there also sufficient information to relate internal dose to risk?

7.3.2 Is There Sufficient Information on the Handling of the Substance by the Body to Justify Biological Monitoring?

After the goals of the programme are defined, it must be decided what sample will be collected to meet them. Information on absorption, distribution, excretion, toxicokinetics, toxicodynamics, and metabolism must all be taken into account to decide on an appropriate sample to collect. It must be established that there is an analyte in the specimen (typically urine, blood, or serum) that correlates with exposure or, if desired, risk. The rate of appearance and disappearance of the analyte in the specimen must also be taken into account in order to decide the timing of the collection. There is no point in collecting an end-of-shift urine sample for a metabolite that only appears in the urine at a later time, nor in delaying until the next day blood sampling for a substance that is cleared from the circulation with a half-life of minutes. Establishing correlations between exposures and the handling of substances or metabolites requires a good deal of scientific effort and is frequently a major impediment to a new monitoring programme.

7.3.3 Is There a Reliable Analytical Method for Measuring the Chosen Parameter?

An appropriate analytical method must be chosen to measure the intended analyte at concentrations that can be expected to be significant for the monitoring programme. It is also desirable to be able to measure background levels of the analyte (if present) in a reference population not exposed occupationally to the substance, so that the spectrum of exposures can be appreciated. Acceptable protocols frequently involve relatively sophisticated technology such as gas–liquid chromatography, gas chromatography–mass spectrometry, or Zeeman-corrected electrothermal atomic absorption. In these cases, adequate monitoring may necessitate the participation of an experienced analytical laboratory. When available for a given analyte, interlaboratory quality assurance programmes are indispensible for ensuring reliable results. The development of new biomarkers will rely increasingly on biotechnologies for which quality assurance programmes are generally unavailable.

7.3.4 Is the Measurement Interpretable?

Even if a correlation between measurement of a substance and exposure or risk has been established in a study population, sources of variation in a particular work force must be appreciated. Causes of individual variation and non-occupational exposure must be considered. Sometimes these will be unimportant, as for example when the only source of a unique metabolite of a pesticide residue would be occupational exposure to that pesticide. In other cases, for instance with many trace metals, other environmental and non-occupational exposures may contribute significantly to the measured value.

7.3.5 Are the Consequences of the Measurement Foreseeable?

Before monitoring is begun, thought must be given to how the results will be used. Can it be decided when an exposure is inappropriate? For many substances there are published guidelines or legislated values which are not to be exceeded. If there are not, a decision must be made on whether to monitor and if so what to set as an action level based on the best knowledge available at the time. A plan must be in place for correcting operational practices if predetermined levels are exceeded.

The reader will rightly conclude that successful monitoring is not always possible, not always necessary, and, when achieved, the culmination of substantial scientific effort. Nevertheless, there is a moral and legal obligation to develop and implement sound programmes for monitoring exposure to harmful substances in the workplace.

7.4 BIOMARKERS AND SENSITIVITY SCREENING

Monitoring exposure to a chemical element by measuring that element in a body compartment, or exposure to an organic compound by measuring an oxidized or conjugated metabolite are conceptually straightforward. However, to get a direct indication of biological effects, or to facilitate monitoring in situations when concentrations of a substance do not correlate with exposures, there is great interest in developing the use of biomarkers. A biomarker may be defined as a parameter used to identify a toxic effect in an organism or an indicator signalling events in a biological system that give a measure of exposure, effect, or susceptibility. Only a few are presently useful, and several more show promise. Several are described here.

7.4.1 Biomarkers of Exposure to Non-carcinogens

Because haemoglobin is plentiful and has a relatively long half-life, it has proven useful in several situations. Carbon monoxide and dichloromethane form carboxyhaemoglobin. Present at about 1% in blood from reference individuals, a value of 3.5% carboxyhaemoglobin is generally taken as the limit of safe exposure to these agents. Some aromatic amines, *e.g.* aniline, generate metabolites that can oxidize $Fe(II)$ in haemoglobin to $Fe(III)$, forming methaemoglobin. Present at less than 2% in the blood of reference individuals, methaemoglobin should not exceed 5%. Certain alkylating agents and pesticides form adducts with amino acids in haemoglobin that can be measured specifically. For example, *N*-aryl pesticides form aromatic amine metabolites that in turn form sulfinic acid amide adducts with cysteine-93 of β-haemoglobin.

Inhibition of enzyme, cell, or tissue function are all possible indicators of adverse exposures. As mentioned above, decreased erythrocyte acetylcholinesterase activity is a well-established means of monitoring exposure to organophosphorus pesticides. Suppression of natural killer cell activity and cytotoxic T lymphocyte function in chronic organophosphorus exposure are also being studied. A control enzyme in haeme synthesis, δ-aminolevulinic acid dehydratase, is extremely sensitive to lead. Therefore, decreased activity of this enzyme in blood and increased excretion of the natural substrate δ-aminolevulinic acid in urine have both been used to monitor lead exposure. Lead also interferes with later steps in haeme synthesis, and the accumulation of free protoporphyrins or zinc porphyrins in erythrocytes have also been used as indicators of lead exposure, although it should be noted that direct measurement of lead in blood by atomic absorption or voltammetric methods is more sensitive than any of these approaches. Cadmium causes damage to the renal tubules, and increased urinary excretion of $β_2$-microglobulin and retinol-binding protein, early indicators of renal tubular dysfunction, may be good indicators of cadmium exposure.

7.4.2   Biomarkers of Exposure to Carcinogens

One of the most important future uses of biomarkers will be in monitoring for exposure to carcinogenic or mutagenic substances. Such substances may be present themselves in very small quantities, and cumulative genetic damage may be more relevant to assessing the cumulative effects of many structurally distinct compounds. The International Agency for Research on Cancer has indicated six classes of biological tests for assessing exposure to carcinogens. None is yet established for evaluating risk of exposures in humans. The six classes are:

- Mutagenicity of urine. Inconsistent results have been obtained in many instances. Smoking is an important confounding factor as the urine of smokers is mutagenic in the assays.
- Thioethers in urine. Many electrophilic genotoxins form glutathione conjugates that are converted to thioethers and excreted in urine.
- Alkyl adducts of DNA or proteins. Many carcinogenic substances are alkylating agents. Alkylated nucleic acids are excised from DNA and excreted in urine, while haemoglobin is a useful indicator of cumulative protein modification. Examples include the measurement of 3-methyladenine or 7-methylguanine in urine after exposure to methylating agents, or *N*-3-hydroxyethylhistidine in haemoglobin resulting from ethylene oxide exposure.
- Chromosomal aberrations. DNA can be obtained from peripheral lymphocytes or cells exfoliated from the mouth or bladder. Damage studied includes sister chromatid exchange, single strand breaks, and point mutations affecting the expression or activity of the enzyme hypoxanthine-guanine-phosphoribosyl transferase.
- Sperm analysis. Functional, morphological, and chromosomal studies may all be useful.
- Oncogene studies. This exciting field is in its infancy as a monitoring tool. An example is the induction of an activating mutation in *ras* by exposure to polycyclic aromatic hydrocarbons.

7.4.3 Biomarkers of Susceptibility

Biomarkers also have a role in assessing individual differences in susceptibilities to chemicals in the workplace. There is evidence that people with glucose-6-phosphate dehydrogenase deficiency and therefore with more fragile red cells may be at increased risk of developing anaemia upon exposure to certain industrial chemicals (*e.g.* aromatic amines). Decreased levels of α_1-antitrypsin predispose to developing emphysema. This protease inhibitor protects the connective tissue of the lung from destruction by proteases, and exposure to proteases is increased when noxious stimuli attract inflammatory cells to the lung. Therefore, workers with partial α_1-antitrypsin deficiency who inhale dusts and particulates may be at an increased risk—as are smokers with this deficiency—of developing lung disease. Individual differences in metabolizing chemicals can also determine relative sensitivities. About half the population acetylates aromatic amines more rapidly than the other half, rendering these amines less carcinogenic. Slow acetylators may therefore be at greater risk of developing aromatic-amine-induced bladder cancer. There is also a wide variation in the rate at which blood from different people can detoxify paroxon ('paroxonase activity'), a toxic metabolite of the insecticide parathion. In none of these cases does there yet appear to be sufficient data on the levels of exposure that will unmask the potential increased risk.

7.5   ETHICAL CONSIDERATIONS

Medicolegal aspects of monitoring and exposure control vary among jurisdictions and only certain general ethical concerns are discussed here without specifics. Any human sampling represents an invasion of privacy and requires the consent of the worker to participate in the monitoring programme. The extent to which such consent can be made a condition of employment is a difficult ethical issue. The monitoring test should be made as non-invasive as possible. Urine collection is relatively non-invasive. At the other extreme, few people will consent to tissue biopsy without very strong indications. Minor risks and discomforts associated with procedures of intermediate invasiveness (*e.g.* blood collection, X-ray exposure) must always be considered. It is, however, the employers' responsibility to ensure that any samples collected from workers are meaningful and will allow the goals of the monitoring programme to be met. As a rule, the least invasive procedure that meets the goals of the monitoring programme should be used. Interpretation must be based on sufficient knowledge, and adequate analytical facilities for obtaining meaningful results should be in place. Of course worker privacy and anonymity must be preserved and the samples must only be used for the purpose for which consent has been given.

The decision to add biological to ambient monitoring is at times contentious. The advantages of biological monitoring have been discussed above, but the worker must never be considered a convenient sampling device. Biological monitoring must not be used as a substitute for appropriate measures of exposure control and practices of industrial hygiene.

A frequent concern of labour organizations is whether measuring internal doses known to be associated with health risks will result in job loss. Monitoring programmes should be proactive and preventive. The worker should not be placed in a position where fear of loss of employment interferes with prompt identification of health risks or an honest assessment of adverse effects. The decision to accept a certain level of risk for secure employment is a personal one, but should never be made under duress. The same concern arises in assessing individual sensitivities. Screening for sensitivities that will result in denial of employment is not generally considered acceptable and, as noted above, there is not yet sufficient dose–response information to consider this approach in any case known to this author. The decision to employ women of child-bearing years in jobs where there is a risk of exposure to teratogenic or genotoxic substances is particularly difficult.

7.6   BIBLIOGRAPHY

N.A. Ashford, C.J. Spadafor, D.B. Hattis, and C.C. Caldart, 'Monitoring the Worker for Exposure and Disease: Scientific, Legal, and Ethical Considerations in the Use of Biomarkers'. Johns Hopkins University Press, Baltimore, MD, 1990.

H. Bartsch, K. Hemmink, and I. O'Neill, eds. 'Methods for Detecting DNA Damaging Agents in Humans: Applications in Cancer Epidemiology and Prevention', IARC Scientific Publication No. 89. International Agency for Research on Cancer, Lyons, 1988.

Deutsche Forschungsgemeinschaft, 'MAK- and BAT-Values 1992'. VCH, New York, 1992.

American Conference of Governmental Industrial Hygienists, 'Documentation of the Threshold Limit Values and Biological Exposure Indices'. ACGIH, Cincinnati, OH, 1989.

T.J. Kneip and J.V. Crable, eds. 'Methods for Biological Monitoring: A Manual for Assessing Human Exposure to Hazardous Substances'. American Public Health Association, Washington, DC, 1988.

R.R. Lauwerys and P. Hoet, 'Industrial Chemical Exposure. Guidelines for Biological Monitoring'. Lewis, Boca Raton, FL, 2nd edn, 1993.

M.A. Saleh, J.N. Blancato, and C.H. Nauman, eds. 'Biomarkers of Human Exposure to Pesticides'. American Chemical Society, Washington, DC, 1994.

World Health Organization, Guidelines on Biological Monitoring of Chemical Exposure in the Workplace (vols. 1 and 2). WHO, Geneva, in the press.

CHAPTER 8

Mutagenicity

DOUGLAS McGREGOR

8.1   STRUCTURE OF DNA (DEOXYRIBONUCLEIC ACID)

Mutagenicity is the process by which the information for determining and maintaining the integrity, functions, and relationships of living cells is permanently changed. A mutagen is an agent that damages the inherited information stored in the cell's genetic material. Mutagenic damage may be caused by the absorption of high-energy radiation, reaction with chemicals, or an interaction with an invading virus. Chemical mutagens come either from the environment of the cell or are produced by the cells themselves as part of their normal metabolism. The principal target for damage is DNA, a helical molecule which is primarily organized into either a single filamentous, supercoiled fibre attached at one or more points to the cell membrane, or distributed between a number of chromosomes contained by a nuclear membrane. When present in the former condition, the organisms are called prokaryotes and when found in the latter condition they are called eukaryotes. DNA is not confined to the nuclear chromosomes, but also occurs outside the nucleus, in mitochondria and in the chloroplasts of plants.

In typical eukaryotic cells, each chromosome contains a single, very large, linear DNA molecule, commonly of the order of 10^7 to 10^9 base-pairs in length. It is complexed with special DNA-binding proteins of two types (to form chromatin): histones and non-histone chromosomal proteins. There are five types of histone, all of which are small, highly basic proteins rich in lysine and arginine. These are the structural proteins of chromatin and, on a weight basis, they approximately equal the weight of DNA. Non-histone chromosomal proteins form a bewildering array of around 10^3 different proteins, such as polymerases and other nuclear enzymes, hormone receptor proteins, and many kinds of regulatory proteins.

DNA (Figure 8.1) is a twisted, ladder-like structure, where the 'poles' of the ladder are chains of deoxyriboses in which the 3' position of one deoxyribose is linked to the 5' position of its neighbour on the same 'pole' by phosphates. These two deoxyribose phosphate chains run in opposite directions, so that the 3'-OH of one chain is adjacent to the 5'-OH of the other chain. The 'rungs' of the ladder consist of complementary pairs of pyrimidines [thymine (T) and cytosine

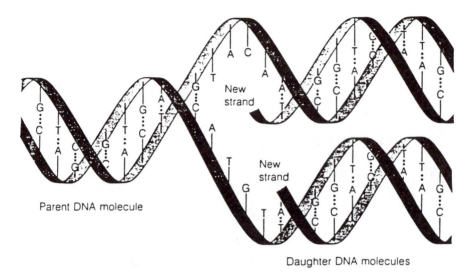

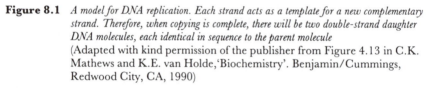

Figure 8.1   *A model for DNA replication. Each strand acts as a template for a new complementary*
strand. Therefore, when copying is complete, there will be two double-strand daughter
DNA molecules, each identical in sequence to the parent molecule
(Adapted with kind permission of the publisher from Figure 4.13 in C.K.
Mathews and K.E. van Holde, 'Biochemistry'. Benjamin/Cummings,
Redwood City, CA, 1990)

(C)] and purines [adenine (A) and guanine (G)]. These bases are always paired so
that A is with T and G is with C, and it is these bases which form the genetic
code. Base-pairing means that the two chains or strands of the helix are
complementary and, hence, permit the same sequences of bases to recur when
the strands separate during DNA replication and a new chain is synthesised using
the existing chain as a template.

The fundamental unit of information in DNA is the codon, or triplet of bases.
Since there are four different bases, 64 combinations of a triplet code are possible.
There are 20 different α-amino acids which are incorporated into proteins, hence
there is sufficient information and some redundancy to code for all of them as
well as the start and stop signals for polypeptide synthesis (Figure 8.2).

The nucleotide sequence in a gene determines the amino acid sequence in the
protein for which it codes. In prokaryotes there is a direct correspondence
between the nucleotide sequence, the messenger RNA (mRNA) that is transcribed
from it, and the polypeptide chain that is translated from the mRNA. In
eukaryotes, however, there exist regions of the DNA sequence that are never
expressed in a polypeptide chain. These non-coding regions are called introns
and they alternate with the regions that are expressed, the exons.

DNA has a special need for metabolic stability because its information content
must be transmitted virtually intact from one cell to another during cell
replication or reproduction of an organism. This stability is maintained in two
ways. The first includes mechanisms that ensure high replication accuracy, such

SECOND POSITION

	U	C	A	G	
U	Phe	Ser	Tyr	Cys	U
	Phe	Ser	Tyr	Cys	C
	Leu	Ser	Stop	Stop	A
	Leu	Ser	Stop	Trp	G
C	Leu	Pro	His	Arg	U
	Leu	Pro	His	Arg	C
	Leu	Pro	Gln	Arg	A
	Leu	Pro	Gln	Arg	G
A	Ile	Thr	Asn	Ser	U
	Ile	Thr	Asn	Ser	C
	Ile	Thr	Lys	Arg	A
	Met	Thr	Lys	Arg	G
G	Val	Ala	Asp	Gly	U
	Val	Ala	Asp	Gly	C
	Val	Ala	Glu	Gly	A
	Val	Ala	Glu	Gly	G

FIRST POSITION (5′ end)　　THIRD POSITION (3′ end)

Figure 8.2 *The genetic code. The table is arranged so that it is possible to find quickly any amino acid from the three letters (written in the 5′→3′ direction) of the codon. Phe phenylalanine, Leu leucine, Ile isoleucine, Met methianine, Val valine, Ser serine, Pro proline, Thr threonine, Ala alanine, Tyr tyrosine, His histidine, Gln glutamine, Asn asparagine, Lys lysine, Asp aspartic acid, Glu glutamic acid, Cys cysteine, Trp tryptophan, Arg arginine, Ser serine, Gly glycine*
(Adapted with kind permission from Figure 5.16 in C.K. Mathews and K.E. van Holde, 'Biochemistry'. Benjamin/Cummings, Redwood City, CA, 1990)

as 3′-exonucleolytic proofreading, which corrects errors made by DNA polymerases, and the uracil-DNA *N*-glycosylase pathway, which prevents mutations that might result from deamination of cytosine to uracil in DNA. The second way is by mechanisms for repairing genetic information when DNA suffers damage. This damage can be caused by replicative errors that are not corrected or by environmental damage. The latter can result from chemical modification of nucleotides or from photochemical changes following the absorption of high-energy radiation.

8.2　DAMAGE TO DNA

Exposure of DNA to UV radiation of about 260 nm produces a number of photoproducts, prominent amongst which are intrastrand dimers joined by a cyclobutane structure involving carbon atoms 5 and 6 of thymidine. These can be completely removed by a photoreactivating enzyme which, in the presence of light (especially 370 nm), binds to the cyclobutane region of DNA. However, another intrastrand dimer, the 6-4 photoproduct, is not repaired in this way and is the cause of mutations when the DNA replicates. Oxidative damage to DNA within cells is a common result of oxidative metabolism of xenobiotics during which molecular oxygen is reduced in a series of four one-electron steps through superoxide anion and hydrogen peroxide to the hydroxy radical and hydroxy ion. The hydroxy radical in particular is highly reactive and represents the most active mutagen generated by ionizing radiation.

Greater general toxicological importance attaches to the alkylation of DNA (commonly by methylating and ethylating agents) which can cause many base modifications. The target bases are primarily purines (although phosphate oxygen is also a target) and while the N^7 position is frequently the quantitatively dominant site for alkylated products, O^6-alkylguanine is the most mutagenic because this interferes with the normal hydrogen bonding of G with C and there is a very high probability of G with T pairing when the modified strand replicates. Should this occur, there is said to have been a GC $\rightarrow$ AT transition mutation. Many of these alkylations (particularly methyl, but also ethyl and hydroxyethyl) can be removed by the protein O^6-methylguanine-DNA methyltransferase (MGMT), which is capable of functioning as an 'enzyme' only once, since it is inactivated when the alkyl group is transferred from the guanine to a cysteine on the protein. Haloethylation can be particularly damaging to cells unless removed by MGMT, since interstrand cross-links can be formed which prevent strand separation during DNA replication.

Many, much larger molecules can also form adducts with the bases of DNA, including polycyclic aromatic hydrocarbons (produced during the burning of carbonaceous materials), aromatic amines, and heterocyclic aromatic amines (*e.g.* produced from certain amino acids during cooking). Many of these are not immediately reactive with DNA but must first be metabolized, frequently by oxidative systems, to electrophilic intermediates. This metabolism provides a large number of products, but, as in the case of benzo[*a*]pyrene, which is a carcinogenic and mutagenic constituent of any burnt carbonaceous material, there may be a high degree of specificity in the metabolite which seems to be responsible for the carcinogenic/mutagenic effects *in vivo* (see Figures 9.2 and 9.3, pp. 96, 97).

8.3   REPAIR TO DNA

DNA damaged in these various ways is frequently repaired by processes in addition to those already mentioned. These processes have been largely studied in prokaryotes, and while excision repair is also important in mammalian cells, the status of SOS* repair and error-prone repair in mammalian cells is not yet clearly defined.

---

* SOS is an error-prone repair system that helps the cell to save itself in the presence of potentially lethal stresses.

8.3.1 Excision Repair

Excision repair can be initiated by DNA alkylation and arylation, the production of pyrimidine dimers by UV radiation, or the production of apurinic or pyrimidinic sites. Apurinic sites can arise as a result of glycosylase activity removing abnormal bases, or non-enzymic depurination following labilization of the glycosyl linkage due to alkylation on N^3 or N^7 positions. The excision repair enzyme system can also repair cross-linking damage.

There are two main modes of DNA excision repair: 'short-patch' or apurinic repair, and 'long-patch' or nucleotide excision repair. Short-patch repair involves the removal and replacement of only a few (perhaps three or four) nucleotides. Long-patch repair results in the removal of the DNA damaged site and up to about 100 adjacent nucleotides. It is inititated by damage which produces large distortions in the double helix, *e.g.* pyrimidine dimers or adducts with large ring systems (polycyclic aromatic amines or hydrocarbons).

Excision repair can occur throughout the cell-cycle. In *Escherichia coli*, the uvrA protein detects a distortion due to a dimer or some bulky adduct and binds to DNA distant from the damaged site. The uvrB protein binds to the DNA–uvrA protein complex and, by DNA gyrase activity, unwinds the DNA strands down to the damaged area. UvrC, with endonuclease activity, cleaves the damaged DNA strand on either side of the lesion. The damaged segment is unwound from the undamaged strand by helicase II (uvrD gene product), after which DNA polymerase 1 fills the gap and DNA ligase seals the remaining nick.

8.3.2 Post-replication Repair

Post-replication DNA repair is an error-prone mechanism which occurs only during the DNA synthetic phase (S-phase) of the cell-cycle. During DNA replication, if the polymerase enzyme encounters a large site of damage on the template strand, then that portion of DNA cannot be used as a template. The result, then, is a newly synthesized DNA that contains gaps of up to 1000 nucleotides, as well as the possibility of shorter gaps. These gaps and nicks are eventually filled by chain elongation (*i.e.* post-replication repair) during S-phase and are ligated. Because the repair polymerase must use a damaged DNA template, it is probable that the newly synthesized DNA strand will contain errors.

If the repair is not error-free, provided that the cell survives and replicates, then the genetic information can be altered. Although these changes are essential to the evolutionary process leading to the diversity of life, the great majority of them will be harmful to the cell. The severity of the harm depends, to some exent, upon whether the affected cell is a unicellular organism or one of many in a multicellular organism. If the affected cell of the latter is a germ cell that becomes involved in the reproductive process, then the resulting offspring will carry in their genetic material (although it is not necessarily expressed) the potentially harmful information. If the affected cell in a multicellular organism is a somatic cell (*i.e.* one which does not give rise to either spermatozoa or ova), then it may experience impaired function or impaired susceptibility to homeostatic, regulatory controls. In this case the cell may have taken a step along the path leading to the emergence of a cancer.

8.3.3   Base Replacement

Replacement of one base by another has several possible consequences. There may be no effect at all either because the base change is in an intron or because it results in a new codon that codes for the same amino acid, a result of the redundancy in the DNA triplet code mentioned above. Alternatively, base change may result in the coding of a different amino acid. This type of change is a mis-sense mutation. Occasionally, the codon for an amino acid residue within the original polypeptide will be changed to a stop codon. This is a nonsense mutation which results in a truncated polypeptide. A mutation in a pre-existing stop codon may code for another amino acid and the continuation of translation into a polypeptide which is elongated up to the next stop codon.

8.3.4   Deletions and Insertions

Deletions or insertions of bases in a gene may be large or small. If large and in a codon region the almost inevitable result is the prevention of the production of a useful polypeptide. The effects of short deletions or insertions depend upon whether or not they involve multiples of three bases. If whole codons are involved, then the consequence is the deletion or insertion of the corresponding number of amino acid residues. The deletion or insertion of any number of bases *other* than a multiple of three causes a shift in the reading frame during translation. Such a frameshift mutation results in a complete change in the amino acid sequence in the C-terminal direction from the point of mutation. Nonsense and frameshift mutations almost always result in the destruction of protein function, so that, if the protein is essential for cell survival, the cell will die.

In addition to the base changes that can result from the interaction of a mutagenic agent with DNA, the development of a discontinuity in the DNA imposes an extra strain on the structure of the chromosome. This can result in lesions that are microscopically visible (*e.g.* at 400 × magnification or greater) after appropriate staining of the chromosomes. Such lesions are particularly prone to develop if the DNA damage involves DNA interstrand cross-linking or DNA–protein cross-linking agents. Numerical changes in the chromosomes can also occur if there has been interference in their movements during cell division by agents that have formed adducts with the cytoskeletal proteins, particularly (but not only) tubulin. What a cytogeneticist refers to as a chromosome consists of two chromatids, one having been derived from each sex cell when they come together during sexual reproduction to form the zygote. Each pair of chromatids is joined by the centromere, which is also the point of attachment of the cytoskeletal proteins which push and pull the chromosomes during cell division. Each of these chromatids is what has been referred to as a chromosome elsewhere in this and in other texts.

8.4 CHROMOSOMAL CHANGE

Chromosomal damage is defined as a microscopically visible modification of the number or structure of chromosomes. Variations in the number of chromosomes may involve the complete complement (polyploidy) or some of the chromosomes (aneuploidy). The loss of a chromosome is a lethal event, but the gain of a chromosome can be viable and create significant genetic unbalances. Structural changes are mainly the result of breaks in the chromatid arms (Figure 8.3). Some of these are unstable and are not transmitted through successive cellular generations, *e.g.* achromatic gaps, breaks of one or both chromatid arms, chromatid interchanges, acentric fragments, ring, and dicentric chromosomes. Stable structural modifications that are transmissible are inversions, translocations, and some small deletions. These genetic factors clearly play important rôles in many human diseases.

Nevertheless, there has been no demonstration that any agent has increased the frequency of any human mutation that can be transmitted from one generation to the next. Furthermore, those agents that do induce heritable mutations in experimental animals will already be considered to be hazardous

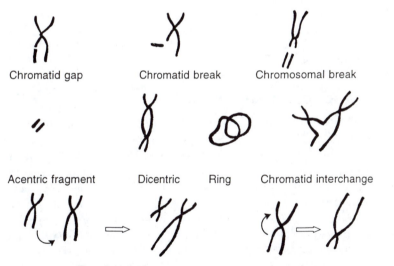

Figure 8.3 *Different categories of chromosomal damage*

chemicals because of their other toxicological properties which are more easily monitored (usually because they are carcinogenic). It also appears that there are no mutagens or clastogens (agents that break chromosomes) that are uniquely active in germ line cells through which their adverse effects can be passed from generation to generation. Therefore, from the viewpoint of toxicological evaluation, if carcinogens are being adequately controlled, then so are mutagens/

clastogens with heritable effects. The main toxicological interest in mutagenicity therefore comes down to what mutagenic activity can tell us about carcinogenicity. Indeed, the prediction of carcinogenicity was the objective that led to the development of mutagenicity assays, and their validation has always been against the yardstick of carcinogenicity.

8.5   MUTAGENICITY AND CLASTOGENICITY ASSAYS

Well over 100 mutagenicity/clastogenicity assays have been developed, but few have been thoroughly validated and only a few will be briefly described here: (a) bacterial cell mutation assay, (b) unscheduled DNA synthesis assays in primary cultures of hepatocytes and (c) the rodent bone-marrow cell micronucleus assay. Protocols vary widely, so the following description will be limited to the principles and main features of the assays. In all cases, there seems to be a trade-off between assay sensitivity and specificity. Here, high specificity has been the hallmark for assay selection, but these assays are also to be found in many published, regulatory guidelines. Nevertheless, the other assays may serve particular purposes and it could also be that some of them have high specificity, but they have yet to be adequately investigated or validated. Examples of other commonly used assays include: yeast and fungal assays for mutation and recombinational events; insect assays (usually with *Drosophila melanogaster*) for mutation and clastogenicity; cultured mammalian cell assays for mutation and clastogenicity; *in vivo* mammalian (including human) assays for clastogenicity, mutation and sister-chromatid exchange.

8.5.1   Bacterial Cell Mutation Assay

By far the most widely used mutagenicity test is the bacteria/microsome assay, more commonly known under the less general name as the Salmonella assay or Ames' assay. This is a gene mutation assay which uses specially constructed bacterial strains, each having a particular type of mutation in an essential gene. The mutations are in the biosynthesis genes for histidine in *Salmonella typhimurium* and for tryptophan in *Escherichia coli*, so the bacteria will grow only if their growth medium contains these amino acids. The mutations consist of base substitutions and frameshifts and the genes can be reverted to the competent coding sequence by the introduction of similar classes of mutations. Hence, it is necessary to use a number of bacterial strains in order to test for and observe different types of mutagenic activity.

Other modifications to the bacteria are designed to increase sensitivity. A cell surface lipopolysaccharide mutation in which the polysaccharide moiety is deleted increases the cell permeability to many hydrophobic substances, besides rendering these organisms much safer to handle. Mutations in various DNA repair genes (notably uvrB) results in an inability to carry out excision repair. Several strains also carry a plasmid which increases sensitivity, probably by introducing an error-prone repair system.

The other important component of the assay is a metabolic activation system, which consists of the $9000 \times g$ supernatant fraction (S9) of homogenized liver (commonly from rats treated with chemicals which induce certain isozymes of the cytochrome P450 complex) and the cofactors required for cytochrome P450 oxidative activity. Essentially, the assay is conducted by mixing a sample of one of the bacterial strains with S9 plus cofactors and the test chemical at a range of concentrations in a suitable vehicle with a dilute ('soft') agar-containing buffer, nutrient salts, glucose, and a small quantity of the essential amino acid coded for by the target gene. This small quantity of the growth-restricting amino acid allows a few rounds of cell replication to occur from all cells surviving the treatment. This complex mixture is poured into a Petri dish, in which the agar is allowed to set, and then incubated for 2 to 3 days at 37 °C. At the end of this period, microscopic examination will reveal a very large number of microcolonies which are not individually visible to the unaided eye, but give the plates a misty appearance. These are the result of the growth permitted by the added histidine or tryptophan. In addition, much larger bacterial colonies will be visible to the unaided eye which arise from mutations occurring during the experiment. These are counted and treatment and control Petri dish colony numbers are compared.

It should be obvious from this description that the test is quick and it is also said to be simple. However, there are many complexities in its successful conduct that require careful control and verification. The results are numerical, but are usually reduced to positive or negative conclusions, since the magnitude of a response can be misleading if comparisons between chemicals are made.

8.5.2 Unscheduled DNA Synthesis

Unscheduled DNA synthesis (UDS) assay is best conducted in primary cultures of hepatocytes. The principle is that DNA damage is repaired and this involves the incorporation of thymidine into the DNA. If the thymidine is labelled with tritium, then the β-tracks in a photographic emulsion can be counted as silver grains. A limitation of the assay is that 'short-patch' repair (see above) is not observed. Normally, hepatocytes are derived from the liver of small rodents, but hepatocytes can be used from any species, including man, provided that they can be obtained in good condition. The cells are allowed to settle and attached to culture vessels and then they are treated with various concentrations of the test substance in the presence of tritiated thymidine for up to about 16 h. The cultures are then washed, fixed, and overlaid with photographic emulsion, which is later developed.

Before examination, the cultures are lightly counter-stained and coded. This examination is for silver-grain counting, which can be either manual or semi-automatic, using a particle counter linked to the microscope by a video camera. Cells in S-phase are not evaluated (these are normally readily recognized by the very dense accumulation of silver grains over the nucleus). For cells not in S-phase (*i.e.* DNA synthesis is not scheduled or part of the cell cycle), silver grains overlying the nucleus are counted and then silver grains over three similar areas of the cytoplasm are counted and averaged to give the 'background'. In fact, in untreated cultures, the background is greater than the nuclear count because of thymidine incorporation into mitochondrial DNA, the synthesis of which is not synchronized with the cell-cycle. Thus, the net nuclear grain count (*i.e.* nucleus minus cytoplasm) is always a negative number in untreated cultures. A significant increase in this number (by whatever criteria are described in the experimental protocol) indicates that UDS has occurred, and this is indicative of DNA damage caused by the treatment.

8.5.3 Rodent Bone-Marrow Cell Micronuclear Assay

Analysis of cells for chromosomal aberrations is a highly skilled and time-consuming activity. However, the micronucleus test is an approximation to cytogenetic analysis which has been developed and used for many years. It can be conducted *in vitro*, but its common utilization is as an *in vivo* assay for chromosomal breakage in bone-marrow cells, mainly of mice. In the development of red blood cells in mammals, the final cell division is followed by the extrusion of the nucleus from the cell. In cells where this was a recent event, the cytoplasm lightly colours blue with the basic stain which would usually be a nuclear stain. This distinguishes the polychromatic erythrocytes from older normochromatic erythrocytes. However, a fragment of a broken chromosome, or even a whole chromosome which lags behind the others when they are being pushed and pulled towards the poles of the cell during the final division, may find that it is not enclosed within the newly formed nuclear membrane. This lagging could also happen because the fragment does not possess the centromere, to which the cytoskeletal apparatus attaches; it could happen to one or more chromosomes if the cytoskeletal proteins (tubulin in particular) have been damaged by protein adducts. These fragments of chromatin within the polychromatic erythrocytes stain in the same way as any chromatin and appear to be micronuclei.

The procedure, then, is to count polychromatic erythrocytes (1000 or 2000 cells per animal are commonly assessed) and score those which contain micronuclei in smears from the bone-marrow of mice treated (usually 24 h) previously with high doses [up to $0.8 \times LD_{50}$(median lethal dose)] of the test compound administered by an appropriate route. Comparison of control with treatment groups will indicate whether there has been an increase in certain types of chromosomal aberrations. While rearrangements are genetically important, they are also much rarer events than simple breaks, so the latter are the dominant contributors to the total damage in a standard cytogenetic analysis and most micronuclei consist of chromosomal fragments. Sometimes a study will allow for the distinction of centromeric material in micronuclei. An increase in such material would indicate that aneuploidy had occurred.

When evaluating the results from these tests which have just been outlined, it is advisable to keep clearly in mind why they were done and the limitations of the data. Their main purpose, it has been argued, is either to provide some information on the possible mode of action of a carcinogen, or to predict whether or not a particular chemical is a carcinogen. In the latter context, a positive response suggests that there is a high probability that the chemical is indeed a carcinogen, as defined by current rodent bioassay testing standards, but a negative result tells us nothing more certain than could be obtained by tossing a coin. It should not be forgotten, however, that even the predictions suggested by a positive result only apply to a population of chemicals and so there must always remain doubts about the status of any single chemical. This caveat can serve as a useful moderator of our enthusiasm for a particular course of action whenever an extrapolation from one situation or species to another must be applied.

8.6 BIBLIOGRAPHY

J. Ashby, J.M. Gentile, K. Sankaranarayanan, and B.W. Glickman, Report of the international workshop on standardization of genotoxicity procedures, *Mutat. Res.*, 1994, **312**, 195–318.

B.J. Kilbey, M. Legator, W. Nichols, and C. Ramel, eds. 'Handbook of Mutagenicity Test Procedures'. Elsevier, Amsterdam, 2nd edn, 1984.

D.J. Kirkland, D.G. Gatehouse, D. Scott, J. Cole, and M. Richold, eds. 'Basic Mutagenicity Tests: UKEMS Recommended Procedures'. Cambridge University Press, Cambridge, 1990.

I.R. Kirsch. 'The Causes and Consequences of Chromosomal Aberrations'. CRC Press, Boca Raton, FL, 1993.

CHAPTER 9

Carcinogenicity

DOUGLAS McGREGOR

9.1   INTRODUCTION

Tumours are, literally, swellings. They have, however, come to mean a special kind of swelling which, popularly, is called a 'growth' or, if it is a particularly dangerous growth, a 'cancer'. These tentative definitions are made with deference to a highly respected German pathologist, Rudolf Virchow (1821–1902), who once remarked that no man, even under torture, can say exactly what a tumour is. Nevertheless, a workable definition is that a tumour is a tissue mass formed as a result of abnormal, excessive, and inappropriate cell proliferation, the growth of which continues indefinitely and regardless of the mechanisms that control normal cellular proliferation.

Nineteenth century pathologists divided tumours into benign, simple, or innocent tumours and malignant tumours. Benign tumours (which can still be dangerous in some situations) remain localized, forming a single mass that is often symptomless and can usually be excised completely. If symptoms are experienced, these are usually due to pressure on adjacent tissues or to excessive hormone production. Malignant tumours invade adjacent tissues and their cells may dissociate and spread through blood and lymphatic vessels to other parts of the body, where they may lodge, divide, and give rise to secondary tumours or metastases. These are difficult to control and account for most of the approximately 20% of the deaths due to cancer in most developed countries.

Cancer as a cause of death is not negligible at any age, but it is primarily a terminal illness in aged populations. Carcinogenesis is the process involved in the development of malignant tumours. A carcinogen, therefore, is a risk factor causally related to an increase in cancer incidence or prevalence. Cancer epidemiologists study the factors by which people develop or die of cancer. To an experimental oncologist (someone who performs experiments in his or her studies of any kind of tumour), however, a carcinogen is not necessarily an agent causing a pathological phenomenon resulting in death. A carcinogen in this context is an agent that increases the incidence of any neoplasm (a new growth), a term applied to all types of tumour, irrespective of whether it is lethal or potentially lethal. In extreme circumstances, a carcinogen may be defined also as an agent that increases the incidence of certain types of pre-neoplastic change.

Carcinogens may be physical, chemical, or biological agents. Important physical phenomena that are carcinogenic are UV and ionizing radiation. Circumstances involving biological agents that are judged to be carcinogenic are chronic infection with the hepatitis viruses B and C, and infection with the blood parasite, *Schistosoma haematobium,* and the liver fluke, *Opisthorchis viverrini.* However, there are many more chemicals that have been recognised as human carcinogens and some circumstances that entail exposures to carcinogens (without, necessarily, identifying the actual carcinogens). Those identified by the International Agency for Research on Cancer (IARC) as of May 1996 are listed in Table 9.1.

9.2 CARCINOGENIC TESTING

The currently accepted design for carcinogenicity testing is to expose rats and mice to the agent for at least 2 years. Rarely is any other species used. For each species, 50 male and 50 female animals per group are dosed with a vehicle or the test agent in that vehicle. Daily observations are made and if any animals become moribund during the experiment, then they are killed so that tissues are not lost through autolysis. All animals, including those that survive to the scheduled end of the experiment, are subjected to autopsy and almost 40 different tissues are taken from each animal for histological examination. All observations are recorded, summarized, and analysed. While this is the currently accepted basic design for regulatory purposes, it has not always been so, and special experimental designs may be used in particular circumstances. If properly justified, these also may be acceptable to regulatory authorities.

The objective is to expose a statistically acceptable number of animals to the highest dose that they will tolerate without reducing their life span for reasons other than tumour development. A series of preliminary toxicity tests is conducted, the purpose of which is to identify a minimally toxic dose, as manifested by reduced body weight gain, reduced food intake, altered appearance, or altered behaviour patterns, and non-neoplastic histology observed at the end of the experiment. This is the so-called maximum tolerated dose (MTD) determination. There are many ways in which the MTD can and has been defined, but the most commonly used definition states: the MTD is the highest dose that can be predicted not to alter the animals' normal longevity from effects other than carcinogenicity. In practical terms the MTD is the dose which, in a 3 month preliminary study, causes no more than a 10% weight decrement as compared with the appropriate control group and does not produce mortality, clinical signs of toxicity, or pathological lesions (other than those that may be related to a neoplastic response) that would be predicted to shorten an animal's natural life span. Ideally, this is what should happen in the carcinogenicity experiment, but in reality the prediction may be incorrect for various reasons. In the carcinogenicity test, animals may die too early, thereby invalidating the study, or the predicted toxicity may not be realized, which invites the criticism that higher dose levels could have been used.

Table 9.1   *Chemical and physical agents and exposure circumstances identified as human carcinogens by the International Agency for Research on Cancer.*

Agents and groups of agents	Mixtures
Aflatoxins	Alcoholic beverages
4-Aminobiphenyl	Analgesic mixtures containing phenacetin
Arsenic and arsenic compounds[1]	Betel quid with tobacco
Asbestos	Coal-tar pitches
Azathioprine	Coal-tars
Benzene	Mineral oils, untreated and mildly treated
Benzidine	Salted fish (Chinese style)
Beryllium and beryllium compounds[2]	Shale-oils
N,N-Bis(2-chloroethyl)-2-naphthylamine (Chlornaphazine)	Soots
Bis (chloromethyl) ether and chloromethyl methyl ether (technical grade)	Tobacco products, smokeless (chewing tobacco, oral snuff)
1,4-Butanediol dimethanesulfonate (Myleran)	Tobacco smoke
Cadmium and cadmium compounds[2]	Wood dust
Chlorambucil	
1-(2-Chloroethyl)-3-(4-methylcyclohexyl)-1-nitrosourea (methyl-CCNU)	*Exposure circumstances*
Chromium(VI) compounds[2]	Aluminium production
Cyclophosphamide	Auramine, manufacture of
Cyclosporin	Boot and shoe manufacture and repair
Diethylstilboestrol	Coal gasification
Erionite	Coke production
Ethylene oxide	Furniture and cabinet making
Melphalan	Haematite mining (underground) with exposure to radon
8-Methoxypsoralen (methoxsalen) plus ultraviolet A radiation	Iron and steel founding
MOPP and other combined chemotherapy, including alkylating agents	Isopropanol manufacture (strong-acid process)
Mustard gas (sulfur mustard)	Magenta, manufacture of
2-Naphthylamine	Painter (occupational exposure as a)
Nickel compounds[2]	Rubber industry
Oestrogen replacement therapy	Strong-inorganic-acid mists containing sulfuric acid (occupational exposure to)
Oestrogens, non-steroidal[1]	
Oestrogens, steroidal[1]	
Oral contraceptives, combined[3]	
Oral contraceptives, sequential	
Radon and its decay products	
Solar radiation	
Talc containing asbestiform fibres	
Tamoxifen[4]	
Thiotepa	
Treosulphan	
Vinyl chloride	

[1] This evaluation applies to the group of chemicals as a whole and not necessarily to all individual chemicals within the group.

[2] Evaluated as a group.

[3] There is also conclusive evidence that these agents have a protective effect against cancers of the ovary and endometrium.

[4] There is also conclusive evidence that tamoxifen reduces the risk of contralateral breast cancer in women with a previous diagnosis of breast cancer.

Other considerations in establishing the MTD value are metabolism and pharmacokinetics. Disproportionate changes in these parameters with increasing dose may signal saturation of metabolic pathways that are dominant at lower, and therefore more probably encountered, human dose levels.

The MTD level is clearly needed for the experimental demonstration of a carcinogenic response in many instances. The basic problem is that the rodent carcinogenicity bioassay is insensitive: in comparison with human populations, the numbers of individuals that are exposed are low. Hence, it is readily conceded that some means must be used to compensate for the low statistical power inherent in the assay and that the use of high dose levels may be one way of compensating for this weakness. For those substances that are known to be human carcinogens, the rodent carcinogenicity tests may be able to demonstrate their potential at doses lower than the MTD. For a sample of 13 human carcinogens for which there were also experimental carcinogenicity data, the rodent carcinogenicity results indicated that the lowest observed effective doses ranged from 0.005 x MTD to 0.5 x MTD. It would be wrong to conclude, however, that a substance showing a carcinogenic effect only at the MTD in the rodent bioassay is not a human carcinogen; these compounds may not be recognizable human carcinogens only for reasons of limited statistical power in the epidemiological studies, or the absence of significant human exposure. At the same time, however, the use of such high dose levels does raise legitimate concerns regarding the mechanisms and relevance of any tumour induction in these assays.

9.3   EPIDEMIOLOGY

Epidemiology also has its complexities in spite of being able to offer, as its major strength, relevancy of the species studied, thereby avoiding the greatest weakness of the experimental approach. The two main techniques used are cohort and case-control studies. Although most epidemiology concerns causality, it should be clearly understood that epidemiological evidence by itself may be insufficient to establish causality. Also, some differences between the epidemiological and experimental approaches to carcinogen identification may not be immediately obvious. These arise because of differences in the pathology and data collection. As mentioned above, all the experimental animals are subject to extensive histopathology as well as autopsy. Such detailed study is almost never undertaken on people, although there are national and regional policy differences on this matter. While the experimental histopathology is restricted to a small sample of each of the (up to) 40 tissues taken, that sample is very much more representative of a mouse or rat than it would be of the much larger human body. Thus, there is a higher probability of discovering small, benign neoplasms in the 2 year old rodent, and these are all taken into account in the subsequent analysis. They are not looked for in any but a few human autopsies and epidemiological studies commonly take into account only the neoplasm which was either clinically diagnosed (with or without histological verification), or was the cause of death or contributed to it: *i.e.* these studies are dependent upon what is clinically important on death certificates, whereas the data used for the analysis of experimental studies are collected directly from the experiment and are not filtered according to their clinical significance. Common incidental human tumours (which do not find a mention on death certificates) are tumours of the prostate, thyroid, adrenal, and kidney.

9.4   MECHANISMS OF CARCINOGENICITY

Research into the process of carcinogenicity, rather than identification of carcinogens, has revealed that cancer is a collection of multistep, multifactorial diseases (Figure 9.1). An important step in our progress to this stage in our understanding of carcinogenicity has been the realization of the importance of metabolism as a process which can convert a large number of apparently very different chemical structures into a common form of chemical reactivity. While the metabolism of a chemical may be complex, a frequent step (called an activating reaction) is the generation of an electrophile which can readily react with nucleophilic centres in complex biological molecules, such as nucleic acids and proteins. Probably the most complete data on metabolism and on mechanisms of activation have been collected for benzo[a]pyrene (Figure 9.2). Although metabolism can occur *via* a variety of routes, the activation pathway *in vivo* proceeds as a consequence of oxidation by the cytochrome P450 family of enzymes *via* the 7,8-oxide and 7,8-dihydrodiol to the 7,8-diol 9,10-oxides, the so-called bay-region diol-epoxides, which are generally regarded as the ultimate, carcinogenic, and DNA-reactive species of benzo[a]pyrene. Both the 7,8-oxide and the 7,8-dihydrodiol exist as pairs of enantiomers, while the 7,8-diol 9,10-oxides are formed as two diasteriomers, each comprising a pair of enantiomers (see Figure 9.3.). Of these four diol-epoxides, the (+)anti-7,8-diol 9,10-oxide is the most biologically active and adducts derived from the covalent binding of anti-7,8-diol 9,10-oxides to DNA have been identified as the major DNA adducts following the exposure of cultured cells or tissues to benzo[a]pyrene. It is tumorigenic to mouse skin and to the lung of newborn mice, while the (-)anti-7,8-diol 9,10-oxide is inactive in these systems.

There are several potential consequences of adduct formation, but the important one as far as carcinogenesis is concerned is that the cells containing the adducts survive and there is some misrepair of the molecular lesion resulting in a change in the proliferation/differentiation control of the cells so that a subpopulation develops with an increased susceptibility to neoplasia. The cells are said to have become initiated.

Cell division is necessary at all stages of neoplasia. In the early stages, this process allows the formation of a clone of initiated cells, within which further genetic changes may occur, leading to a yet more susceptible cell population. Out of this promotion phase, histologically recognizable pre-neoplastic lesions emerge. Most of these develop no further, but a small number (perhaps only one) may experience more genetic changes, as a result of which a cell population can arise which is no longer susceptible to the usual cell population size controls. This conversion to autonomous growth marks the emergence of a tumour, but yet

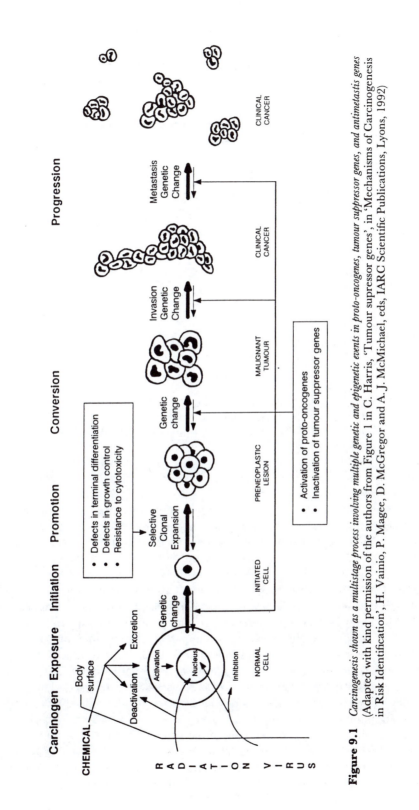

Figure 9.1 *Carcinogenesis shown as a multistage process involving multiple genetic and epigenetic events in proto-oncogenes, tumour suppressor genes, and antimetastasis genes* (Adapted with kind permission of the authors from Figure 1 in C. Harris, 'Tumour suppressor genes', in 'Mechanisms of Carcinogenesis in Risk Identification', H. Vainio, P. Magee, D. McGregor and A. J. McMichael, eds, IARC Scientific Publications, Lyons, 1992)

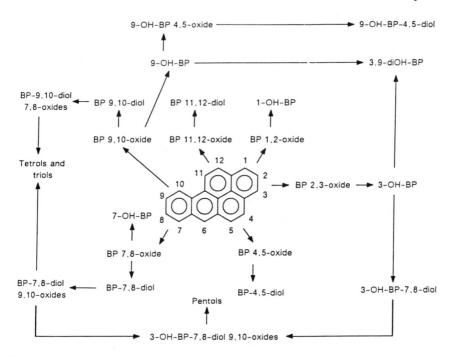

Figure 9.2 *Pathways involved in the metabolism of benzo*[a]*pyrene (BP)*
(Adapted with kind permission of the publisher from Figure 1 in D.H.
Phillips and P.L. Grover, Polycyclic hydrocarbon activation: Bay regions and
beyond, *Drug Metab. Rev.*, 1994, **26**, 443–467)

more changes are necessary if the tumour is going to progress to a spreading type
of neoplasm (metastasis).

The agents that induce the genetic changes and produce the selective pressure
favouring the clonal expansion of cell populations are not necessarily the same
throughout the neoplastic process and may not act directly upon the cells' genetic
material (DNA). It is notable that one of the most potent rodent carcinogens
known (3,4,7,8-tetrachlorodibenzo-*p*-dioxin) shows no sign of being able to
damage DNA in a wide range of assays. This and some other chemical
carcinogens (including certain hormones and hormone-like substances) appear to
act initially through receptor mechanisms. Other compounds which are rodent
carcinogens may also act through so-called non-genotoxic mechanisms, examples
of some proposals being: (a) blood lipid-lowering compounds which induce a
specific proliferation of cytoplasmic organelles, called peroxisomes, in the liver;
(b) compounds which cause an accumulation of a specific protein, $\alpha_{2\mu}$-globulin, in
the kidney of male rats; (c) compounds which interfere with the feedback control
of thyroid, pituitary, and ovarian hormones; (d) diet-enhanced calcium absorption which at least leads to hyperplasia of the adrenal medulla. The importance
of these observations in the mechanism of carcinogenesis is not yet clear: these
particular common properties of the chemicals may have significance ranging
from negligible to substantial importance for their carcinogenicity. However,

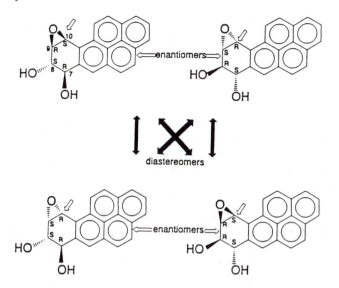

Figure 9.3 *The configurational isomers of the bay region 7,8-dihydrodiol 9,10-oxides of benzo[a]-pyrene*
(Adapted with kind permission of the author from Figure 4 in A. Dipple, 'Reactions of polycyclic hydrocarbons with DNA', in 'DNA Adducts: Identification and Biological Significance', K. Hemminki, A. Dipple, D.E.G. Shuker, F.F. Kadlubar, D. Segerbäck, and H. Bartsch, eds., IARC Scientific Publications, Lyons, 1994)

irrespective of whether a carcinogen has or has not any direct effect upon the cell's DNA, changes in gene expression always occur during carcinogenesis.

9.4.1 Genes

Currently, it is thought that genes of three main types may be involved in carcinogenesis: oncogenes, tumour suppressor genes, and DNA repair enzyme genes. The concept that there are genes capable of causing cancer (oncogenes) is based largely on studies with transplantable tumours in chickens, mice, and rats. In 1911, Rous found that a transmissible sarcoma in chickens could be induced by a cell-free filtrate of the tumour. Similar results were found for some rodent sarcomas and the responsible agent was later demonstrated to be an RNA virus. There is, however, little evidence that any human tumour is induced by an RNA virus (except in the case of HCV, chronic infection with which does increase the risks of liver cancer). In 1965, however, it was proposed that the cells of many, if not all, vertebrates possess the information in their DNA to produce RNA viruses. This information (the virogene) could code for virus, but a portion of it (the oncogene) was responsible for malignant transformation. Normally, this DNA is not expressed, but it could be activated as a result of exposure to chemical carcinogens or radiation. During the following years, it was found that while some transforming RNA viruses carry the information to induce tumours quickly and directly (oncogenes), others induced transformation slowly, apparently by insertion of reverse-transcribed DNA adjacent to host cellular genes, in some way altering the transcription of them.

9.4.2 DNA Analysis

Analysis of DNA from many different organisms, including yeast, insects, birds, and mammals (including man) revealed that it contains base sequences with a high degree of homology to the oncogene region of the viruses. These normal cellular DNA sequences are called proto-oncogenes, or c-*onc* sequences.

It is highly probable that the viral oncogenes (or v-*onc* sequences) are derived from the proto-oncogenes. Well over 20 oncogenes have been identified and, in normal cells, their expression is well controlled. Their functions have been determined in some cases and they appear to play a rôle in growth and development.

The known proto-oncogene products include growth factors (*e.g. sis*), protein kinases (*e.g. src*) which phosphorylate serine, threonine, or tyrosine, including membrane receptors with protein kinase activity; GTP-binding proteins (*e.g. ras*), and nuclear proteins (*e.g. c-myc*) which seem to be involved in the control of cellular proliferation and differentiation. For some proto-oncogenes, there is a close association between their expression and cell proliferation. Stimulation of a non-malignant cell into division often depends upon a signal initiated by an extracellular chemical. This chemical interacts with a receptor on the cell membrane, which transfers the signal through the membrane and into the cytoplasm and, ultimately, to the nucleus where DNA synthesis is initiated. Proto-oncogene products have been found to function at each step of this pathway. Alteration of the proto-oncogenes may result in a development along a pathway leading to neoplasia. Inappropriate activation of a signal transduction pathway (*e.g.* by sustained exposure to certain hormones or exogenous chemicals such as 3,4,7,8-tetrachlorodibenzo-*p*-dioxin) may have a similar effect. Activation may occur by mutation, by amplification, or by rearrangements such as chromosomal translocation. Activated proto-oncogenes (oncogenes) have been detected in malignant cells of human origin by their ability to transform normal cells after transfection of malignant cell DNA and by using labelled DNA probes that are complementary to known viral oncogenes. Nevertheless, analysis of human tumours has resulted in the detection of oncogenes in only about 10% of them, although methodological difficulties may contribute to this low proportion.

Tumour-suppressor genes code for products which control some aspect of normal cell turnover within a tissue. Loss of the effective gene product by mutation of the gene therefore results in reduced control, *e.g.* of cellular proliferation, differentiated state or programmed cellular death (apoptosis). However, only one active gene is necessary for suppression. Since there are normally two genes (except on the sex-chromosomes), both copies (or alleles) need to be lost before tumour suppression is lost. The concept of tumour suppressor genes arose from observations on two human, childhood tumours: retinoblastoma, which affects the eye, and nephroblastoma or Wilms' tumour, which affects the kidney.

9.4.3 Hereditary and Non-hereditary Forms

Retinoblastoma occurs in both hereditary (40%) and non-hereditary forms. The non-hereditary form is always unilateral, whereas about 80% of the hereditary form cases have bilateral disease. In most families carrying the germ line mutation which predisposes to retinoblastoma, the penetrance of the mutation is 90–95% and affects both sexes (*i.e.* it is an autosomal dominant trait). The germ-line mutation has been traced to the long arm (*i.e.* q) of chromosome 13 (*i.e.* 13q) and localized to a particular staining band of the chromosome which is identified as 13q14.11, reading from the centromere (which separates the short, p, arm from q). This locus has been named RB1 (although it seems to be the only locus at which mutation can produce retinoblastoma). The normal cells that are able to transform into retinoblastoma are probably embryonic cells which differentiate into retinoreceptors early in life. Once differentiated, they are no longer at risk of transformation by mutation (because the mature cells no longer divide). For non-hereditary retinoblastoma, both mutations at the RB1 locus are somatic and therefore must occur in the same cell. The probability of two independent mutations occurring in this way is very low indeed. However, when the first mutation is in the germ-line, then every potential target cell in the retina has undergone the first mutation and the probability of the required second mutation occurring in just one cell of the total target population is very much higher, and so bilateral disease much more likely.

Wilms' tumour has many of the characteristics of retinoblastoma. In this case the tumour suppressor function is located on the short arm of chromosome 11 (at 11p13.05-06). A number of other tumour-specific suppressor genes have been identified, but recent attention has been drawn to p53, which is located on chromosome 17q12-13.3, because it seems to be associated with several different tumours, including astrocytomas, osteocytomas, and carcinomas of the breast, colon, liver, oesophagus, and lung. Its behaviour is, however, complex, and much remains to be learned about its true significance.

The importance of deficiencies in DNA repair in carcinogenesis is suggested by the observation of highly elevated cancer incidences in people with well-characterized autosomal, recessive diseases which involve high sensitivity to DNA-damaging agents: xeroderma pigmentosum, Fanconi's anaemia, ataxia telangiectasia, and Bloom's syndrome. Xeroderma pigmentosum carries with it a deficiency in the excision repair of DNA lesions induced by UV irradiation. A very high proportion of such people develop benign and/or malignant skin tumours, particularly if they are exposed to the sun or other sources of UV irradiation.

Fanconi's anaemia patients only inefficiently repair enzyme-induced incision of DNA at cross-links. They have an increased risk of developing liver tumours, skin carcinomas and leukaemias.

Ataxia telangiectasia patients' DNA is highly susceptible to damage by ionizing radiation and chromosome-breaking chemicals. About 10% of them develop malignancies, mainly of the lymphoid system, but another factor contributing to their increased cancer risk is the commonly associated immunodeficiency. Bloom's syndrome patients also show an increased frequency of chromosomal damage and it appears that the disease may result from a mutation in the gene coding for DNA ligase I, which is an important enzyme during DNA replication. Approximately 25% of these patients develop a leukaemia, lymphoma, or carcinoma.

These research areas impact upon our understanding of carcinogenicity as a disease process and may suggest new approaches to therapy, but clarification of their rôle in the story is also important to the way in which changes catalogued in experimental and epidemiological studies should be interpreted in the evaluation process, thereby determining our level of concern when an excess of tumours of a particular type is observed.

9.5   BIBLIOGRAPHY

A. Balmain and K. Brown, Oncogene activation in chemical carcinogenesis, *Adv. Cancer Res.* 1988, **51**, 147–182.

H.C. Grice and J.L. Ciminera, eds. 'Carcinogenicity: The Design, Analysis and Interpretation of Long-term Animal Studies', Springer, New York, 1988.

A.J. Levine, The tumor suppressor genes, *Annu. Rev. Biochem.*, 1993; **62**, 623-651.

H. Vainio, P.N. Magee, D.B. McGregor, and A.J. McMichael, eds., 'Mechanisms of Carcinogenesis in Risk Identification'. IARC Scientific Publication No. 116, IARC Scientific Publications, Lyons, 1992.

B. Vogelstein, E.R. Fearon, S.R. Hamilton, S.E. Kern, A.C. Preisinger, M. Leppert, Y. Nakamura, R. White, A.M. Smits, and J.L. Bos, Genetic alterations during colorectal-tumor development, *N. Engl. J. Med.*, 1988, **319**, 525–532.

CHAPTER 10

Reproductive Toxicity

FRANK M. SULLIVAN

10.1 INTRODUCTION

In recent years there have been differences between countries in the meanings
applied to the terms used in this field and currently there is an initiative by the
World Health Organization and the Organization for Economic Co-operation
and Development (OECD) to attempt to harmonize the terminology used in
reproductive toxicology to agree on the meaning of the main terms used. There is
increasing agreement between the USA and Europe that the term Reproductive
Toxicity should be defined as any adverse effect on any aspect of male or female
sexual structure or function, or on the developing embryo or fetus, or postnatally,
which would interfere with the production or development of a normal offspring
which can be reared to sexual maturity, capable in turn of reproducing the
species. This is a very wide definition and includes various types of toxicity which
are often considered separately. The two main subdivisions of reproductive
toxicity are (i) sexual function and fertility in males and females, and (ii)
developmental toxicity to the embryo and fetus. Sexual function and fertility
refers to effects on the male and female sexual behaviour and gonads. This
includes any effects from puberty to conception and on development of the
fertilized ovum up to the stage of implantation in the uterine wall. Developmental
toxicity includes adverse effects on embryofetal development from the stage of
implantation through parturition and postnatal development up to the stage of
puberty. It may also include adverse effects on lactation which could interfere
with normal postnatal development, either by altering the quality or quantity of
milk produced, or by the passage of chemicals into the milk to affect neonatal
development.

Various other words are used which have specific meanings. For example,
teratogenicity is the ability to cause gross structural (anatomical) malformations in
the developing embryo/fetus; behavioural teratogenicity is a term which has been
used to describe the ability to affect the developing embryo/fetus in such a way as
to result in abnormal nervous system development either to impair the neurological or intellectual development or affect the behaviour of the offspring after
birth. Since a good deal of central nervous system functional and biochemical

development occurs not only *in utero* but also in the early postnatal period up to about 2 years of age in humans, it is possible that chemicals which affect lactation or which are transferred to the infant *via* the milk may also affect the normal nervous system development of the offspring in such a way as to produce permanent effects. The term embryofetal toxicity is commonly used to describe all of the different types of toxicity that may affect the conceptus without defining whether these are induced in the embryonic period or in the fetal period. These include death which is followed in rodents by resorption of the fetal remains, and in primates and humans by abortion. Also included are intrauterine growth retardation with decreased weight and retardation of ossification of the fetuses, as well as an increase in the minor variations which commonly occur in fetuses such as changes in the proportion of fetuses with 12 or 13 or 14 pairs of ribs. The term developmental toxicity is often used now to include all of these different aspects of reproductive toxicity which can result in abnormal structural or functional development of the offspring following exposure of pregnant or lactating females.

10.2 RISK ASSESSMENT FOR REPRODUCTIVE TOXICITY

Risk assessment, as described in Chapter 5, is a process which involves three phases of hazard identification, dose–response extrapolation, and exposure assessment. This then allows an estimate to be made of the risk from a defined exposure. Each one of the phases in the assessment is complex and open to many variations, and provides problems which have to be overcome. Hazard identification usually involves the detection of reproductive toxic effects in animals using screening tests which are described below. The precise characterization of these effects is carried out in detailed animal experiments, and in humans. Dose–response analysis attempts to define the range of doses which cause, or which do not cause, adverse effects. It also involves extrapolation within and between species and the use of modifying factors. Exposure assessment involves examination of the actual levels of a chemical to which humans are exposed and may need to take into account special features of relevance such as route of exposure.

10.3 THRESHOLDS IN REPRODUCTIVE TOXICITY

It is generally accepted in reproductive toxicity that a threshold will exist for each and every adverse effect of chemicals observed, and that exposure to chemicals at levels below the thresholds will not produce any adverse effect on reproduction. In this respect, reproductive toxicity differs fundamentally from carcinogenicity where no thresholds for genotoxic carcinogens are usually accepted. The important consequence of this is to understand that a chemical which may produce gross structural malformations, *i.e.* is teratogenic, at high doses in animals, or which may affect fertility by producing testicular atrophy at high doses in animals, may be completely safe in humans exposed to low doses in the workplace or in the environment. This explains why the observation of adverse effects in animals, *i.e.* hazard identification, has to be coupled to dose–response analysis, in the process of identifying whether any risk will apply to humans exposed in defined circumstances, *i.e.* risk assessment.

10.4 SCREENING TESTS IN ANIMALS FOR REPRODUCTIVE TOXICITY

Hazard identification is normally done by means of animal experiments. Many different designs of study have been used to assess adverse effects of chemicals on fertility and development and these will be discussed in four sections namely, drugs, pesticides, food additives, and industrial chemicals.

10.4.1 Drug Testing

Thalidomide is a hypnotic drug which, if used by pregnant women in the first few weeks of pregnancy, was found to produce severe congenital malformations, usually short or missing limbs, as well as other less obvious defects. One important consequence of its disastrous use in 1957–61, was that virtually every country in the world introduced legal requirements for the safety testing of potential new drugs before these could be released for use by the general public. A major aim was to prevent a repetition of the teratogenic effects of thalidomide, so, not unnaturally, detailed reproductive tests were included as part of the testing battery. In 1966, under the direction of Dr. Lehmann, the US Food and Drug Administration (FDA) published guidelines for a three-segment study for drug testing for adverse effects on fertility and pregnancy. This proved to be a classic design and will be described in detail since it forms the basis of most of the more recent designs introduced worldwide for drugs and other types of chemicals.

10.4.2 Three-Segment Reproduction and Teratogenicity Study

Normally, three dose levels plus controls are used in each segment. Segment 1 of the study is the fertility and general reproductive performance segment and is normally carried out on rats. In this, young male rats are treated for 60–80 days to cover the whole period of spermatogenesis. Female rats are treated for 14 days to cover three estrous cycles and are then mated with the treated males, treatment of the males and females continuing during the mating period. Treatment of the females continues throughout the whole of pregnancy. Half of the females are killed at mid-term for examination for dead and resorbing fetuses, and the remaining females are continued on treatment through parturition and lactation until weaning of the young which is usually at 21 days postpartum. The young are reared till sexual maturity and tested for fertility. The protocol is modified in Europe whereby half the females are killed just before term instead of mid-term so that full fetal examinations may be carried out. In addition, treatment is stopped at parturition, and the pups delivered naturally are tested for any adverse effects on development, such as delay in appearance of landmarks such as development of righting reflexes, auditory and visual function, and development of normal behaviour during the rearing period.

In general however European regulatory authorities accept studies carried out to the USA design. In Japan, there is a major difference in protocol design in that treatment of the mated females is stopped on day 7 of gestation. Otherwise the design is similar. This means that the Japanese version is a simple fertility study, and the possible effects on embryofetal development are studied in a more complex segment 2 study. With all of these designs, if an adverse effect is observed on fertility or pregnancy, then separate mating of treated males with untreated females and *vice versa* may be necessary to demonstrate if the effect is on one sex only, and other studies may be necessary to investigate the exact nature of the effects observed.

Segment 2 of the study is the teratogenicity or embryotoxicity segment and is normally carried out in two species, usually rats and rabbits. Mated animals are treated during the period of organogenesis (days 6–15 in rats, days 6–18 in rabbits) and the fetuses are delivered by Caesarean section on the day before expected parturition (*i.e.* day 21 rats, day 29 rabbits) for examination of the fetuses for malformations. This is necessary in order to prevent loss of deformed fetuses by cannibalism of the affected pups which happens in these species. The uterus and its contents are examined for evidence of dead or resorbing pups and the pups are examined for gross, soft tissue, and skeletal defects by the use of appropriate dissection and staining techniques. In the USA and EEC countries 24 pregnant rats or 12–15 rabbits per dose group are normally used. In Japan, 30 pregnant rats are used per dose group and one-third of them are allowed to deliver the pups naturally. These are reared to weaning and some behavioural and developmental tests carried out before termination at weaning. This Japanese modification gives more information than the original design, especially on whether there is catch-up when developmental retardation is observed in the fetuses, as is often the case. This design is acceptable to most other regulatory authorities and has been incorporated into the most recent international guidelines discussed below.

Segment 3 of the study is the peri-postnatal segment. This is normally carried out on pregnant rats which are treated during the last part of gestation, not covered by treatment in the Segment 2 study, through parturition until weaning. This is to examine whether the drug has any adverse effects on parturition or lactation which have not been detected in the Segment 1 study. Higher doses may be used in this segment than in the Segment 1 study. Problems are commonly seen in the Segment 3 study when non-steroidal anti-inflammatory agents which inhibit prostaglandin synthesis are tested, or when progestagens are tested, since both of these drug classes can markedly prolong gestation in rodents and so lead to parturition difficulties which may result in neonatal deaths. Such effects however do not preclude the use of these drugs in humans in whom such effects would not be allowed to occur.

When positive results are observed in any of the segments in the reproduction studies, these will normally lead to further studies being carried out to investigate the mechanism of action. The effects observed may be extensions of the pharmacological actions of the drug which might be expected at the high doses used in the toxicity tests, but would not be expected at the lower doses used clinically. Sometimes the actions may be exerted on rodent-specific aspects of reproductive physiology and would not be a problem in humans. Differences in results between rats and rabbits may be due to metabolites which are produced in one species but not in another. It is then important to know which metabolites are produced in humans and to be sure that these are adequately tested in the animal studies. It may sometimes be necessary to synthesize adequate quantities of human-specific metabolites and for animal tests to be repeated using these.

10.4.3 International Harmonization of Drug Testing Guidelines

There has been a gradual change in recent years from studies designed mainly to detect teratogenic effects to studies with a much wider range of end-points. In particular there has been increased interest in adverse effects on postnatal development following exposure during pregnancy. This has necessitated modification of the older study designs and, since most drugs are tested for world-wide markets, has resulted in a huge increase in the amount of work and numbers of animals required to satisfy the requirements of the different regulatory authorities. This in turn led to a major revision in 1993 of the drug testing guidelines by the International Conference on Harmonization (ICH, 1993). The new guidelines have been accepted, leading to a reduction in the numbers of animals used, and acceptance of the studies worldwide. The new guidelines are essentially based on the three-segment design, still testing for effects on fertility, embryofetal development and postnatal development, but allow variations in the timing of dosing and various combinations of segments to be used, giving more scope for the toxicologist to produce studies tailored to specific chemicals. Increasing the flexibility of guidelines is regarded by most toxicologists as very desirable, but it does place an extra responsibility on the toxicologist to justify his choice of any particular study design.

Since regulatory requirements for most other types of chemicals, food additives, pesticides, and industrial chemicals have been based on the drug testing guidelines, it is likely that testing of these will change also within the next few years.

10.4.4 Pesticide Testing

The safety evaluation of pesticides for reproductive toxicity has to take into consideration several factors which are not relevant for drug testing. By their nature, pesticides are toxic compounds since their use is to kill target species which may be plants, insects, fungi *etc.*, though they may have a very low toxicity level for mammals. Consideration has to be given to the safety of workers manufacturing and handling these chemicals. Exposure of operators may be considerable and this has to be taken into account in the risk assessment of safety in use. In addition, for chemicals used on food crops, residue data have to be obtained so that exposure of the general public consuming the food can be calculated. For chemicals used on amenity lands such as parks, gardens, and playing fields, consideration has to be given to the exposure of the public using these facilities. The environmental ecological impact must also be examined so that studies may have to be carried out to look for effects on beneficial species such as earthworms and bees, as well as on fishes, birds, predators, and so on. Reproduction tests on these species are outside the scope of this chapter but are of great environmental interest.

For tests related to operator exposure it is common to carry out the Segment 1 and Segment 2 tests described above for drugs. This covers possible effects on male and female fertility and developmental toxicity. For pesticides leaving significant residues on food, it is usual, in addition, to carry out a multigeneration study to test for long-term genetic or cumulative effects or effects on young developing animals. In such tests, male and female animals (usually rats) are treated continuously with the chemical at three dose levels administered in the diet, through several generations. A number of different designs of multigeneration study have been used for pesticides testing. The original design published by the FDA for testing food additives has also been used widely for pesticide testing. This is a three-generation two-litter test, which involves treating young male and female rats with three dose levels of the test substance plus a control group and allowing them to produce two litters, F_1a and F_1b. The pups of the F_1a litter are killed for examination at weaning at 3 weeks of age. Selected pups from the second (F_1b) litter are reared to maturity, still under treatment, and in turn are allowed to produce two litters (F_2a and F_2b). Again, selected pups from the second (F_2b) litter are reared to maturity, still under treatment, and allowed to produce two litters (F_3a and F_3b). The second litter (F_3b) is then subjected to full pathological and histopathological examination for any effects of long-term exposure.

An immense amount of information is obtained in such studies, not only on the different aspects of fertility and pregnancy, but also on the possible effects on growth and development of the young animals. In Europe, in an attempt to reduce the numbers of animals involved without significant loss of information, it is becoming common to perform this test using only two generations with two litters in each. In the USA, a modified test using two generations with one litter in each is acceptable for pesticide testing. Recently, designs have been used where a third litter is produced in one or two of the generations with the dams killed just before delivery for examination of the fetuses, so that the test includes a teratology element as well.

10.4.5 Food Additive Testing

The reproductive toxicity test requirements for the testing of food additives are less well defined in Europe than for the other classes of chemicals considered above. In the USA, however, the FDA have published an extensive account of the types of test which may be required depending on the use, amount, frequency, and duration of expected exposure. In general, fertility and developmental toxicity are tested using Segments 1 and 2 of the drug testing scheme, but sometimes modified so that the period of chemical exposure in the developmental test, is extended to include most of the gestation. Consideration has to be given also to whether the metabolism of the chemicals may alter on chronic administration, so that more than one dosing regimen may have to be used. For food chemicals where there is likely to be prolonged or extensive exposure, multi-generation studies are normally performed using one of the designs discussed above, commonly a two-generation two-litter test.

10.4.6 Industrial Chemicals Testing

In most industrialized countries there are now testing requirements for industrial chemicals when these are produced in significant amounts. In Europe, for example, under the requirements of the laws generally known as the Dangerous Substances Directive there is a stepwise system which permits regulatory authorities to require sequential testing for fertility, teratogenicity, and multigeneration effects as the tonnage of chemical produced per year, or *in toto*, reaches certain critical values. This is described in detail below. The methods used are essentially similar to those described above for drugs. There has, however, been good international co-operation under the auspices of the OECD to establish mutually acceptable reproductive toxicity testing guidelines for chemicals which have acceptance worldwide. As these are regularly updated, interested readers should obtain the most recent versions from the OECD.

10.5 RECENT DEVELOPMENTS IN REPRODUCTIVE TOXICITY TESTS FOR FERTILITY

In females, the traditional methods of assessment of fertility by dosing females for two or three estrous cycles followed by mating trials is still the most common method. This may be supplemented by assessment of the different stages of the oestrous cycles by use of vaginal smears. This gives a good indication of any changes from the normal cyclical hormone activity.

In males, however, there has been a change in attitude in recent years with regard to which methods are the most satisfactory for detecting adverse effects on the testis. In rats, there is normally a large excess of sperm production compared with the number required for fertility to be maintained. It is clear that there can be considerable testicular damage with reduction in testis weight, damage to the seminiferous tubules, reduction in sperm number, and increase in non-motile or abnormal sperm, without there being any detectable change in fertility in the males as assessed by mating performance. This is clearly different from the situation in humans where relatively small reductions in sperm quality may be associated with infertility. Increasing reliance has therefore been placed on histological and sperm quality assessments. This is demonstrated by the changes proposed in the ICH guidelines for drug evaluation and the new OECD guidelines for chemical assessment, which have reduced the time of premating treatment of males from 10 to 4 weeks provided that no histopathology in the testis or epididymis has been observed. Actual mating studies are still required to assess sexual behaviour, and sperm analysis for count, motility, and morphology are commonly performed.

10.6   EXTRAPOLATION OF RESULTS OF ANIMAL STUDIES TO HUMANS

When adverse effects are detected in any of the above screening tests, then extrapolation to humans involves analysis of the mechanisms and sites of action of the chemicals causing the effects. This requires a good knowledge of the comparative physiology of reproduction in the different animal species.

The most satisfactory situation is when one can define the precise mode of action of the reproductive toxicant in the animal model and then state whether such a mechanism would operate in humans at the exposure levels concerned. The differences in reproductive physiology between species are sufficiently large for there to be cases where a chemical may affect fertility or development in rodents or rabbits by acting on hormones or systems which do not operate in humans. In such a situation the animal studies can be discounted, although there is still the possibility of a different effect in humans. If similar mechanisms to those responsible for the adverse effects in animals do exist in humans, then some confidence can be achieved by determining the 'no effect levels' for the underlying changes in human studies where this is possible.

The more usual situation, however, is that the underlying mechanisms of toxicity in animals are not known, and then one has to resort to the use of safety factors in arriving at the acceptable levels of exposure for humans. This is the least scientific part of risk assessment, and is regarded by some as a 'numbers game' and by others as a scientific judgement.

10.7   CLASSIFICATION

10.7.1 The European Community Classification of Chemicals for Reproductive Toxicity

The classification and labelling of dangerous substances was first introduced in 1967 in the European Community with Council Directive 67/548/EEC known as the Dangerous Substances Directive. The Sixth Amendment to this directive in 1979 introduced a notification procedure and a requirement for labelling chemicals for toxicity. Three special categories of labelling were used for carcinogenicity, mutagenicity, and teratogenicity. The teratogenicity classification was restricted to chemicals inducing teratogenic effects in the classical sense of the word, *i.e.* producing only gross structural malformations. Discussions by expert advisors to the European Commission over several years has led to a widening of concern in this area of toxicology, and under the current Seventh Amendment the classification of teratogenicity has been changed and expanded to 'toxic to reproduction'. This includes adverse effects on fertility, pre- and postnatal development, and lactation, and encompasses not only structural malformations but also functional deficits. This has resulted in a major change in the testing requirements to allow adequate classification of chemicals for these other aspects of reproductive toxicity.

10.7.2 Seventh Amendment to EC Directive 67/548/EEC 1992

This Directive relates to the classification, packaging, and labelling of dangerous substances in order to reduce barriers to trade within the Common Market and to approximate the laws of the Member States to permit this. An important function of the Directive is to protect man and the environment, and concerns itself with the health and safety of the population, and especially with the safety of workers exposed to potentially dangerous chemicals. The Directive requires the notification of new chemicals to appropriate authorities for assessment with a view to classifying and labelling them with appropriate health and safety warnings. The definition of dangerous which is used includes the following properties: explosive; oxidizing; flammable; corrosive; toxic including death, acute or chronic damage, irritant or sensitizing, carcinogenic, mutagenic, and toxic for reproduction. The toxicity categories refer to exposure by the inhalation, oral, and dermal routes.

There is also a requirement for production of Safety Data Sheets for chemicals when sold relating to health and safety at work, and manufacturers have to review their own data in order to provide adequate warnings and to obtain toxicity data on their chemicals in cases where such data do not exist at present. As there is also new legislation within the Community and in the USA concerning the health and safety of pregnant women at work, there will be increasing pressure to obtain data on the reproductive hazards of chemicals. Many large companies currently have policies relating to the safety of women workers during pregnancy, but the majority of workplaces are still not prepared to deal with this situation. Most countries have legislation relating to the prevention of unfair sexual discrimination in the workplace and many countries have legislation to protect the rights of the pregnant worker. The converse side of this, however, is the increased risk of litigation in the event of the developing fetus having a malformation which may be claimed to have arisen from exposure of the mother to dangerous chemicals at work.

The effect of the new classification under the Seventh Amendment on the toxicity of chemicals towards reproduction, is exemplified by methyl mercury. Methyl mercury has a severe adverse effect on the central nervous system development, causing blindness, deafness, and mental retardation in the absence of structural defects. This would not have been eligible for classification under the previous legislation but will now be classified under the Seventh Amendment. In addition, the potential to induce adverse effects at any stage of the reproductive cycle, in males as well as females, will now result in classification.

10.7.3 Testing Requirements under the Seventh Amendment

When new substances, or mixtures, are notified under this Directive, the amount of testing required to be submitted to the appropriate national authorities depends on the amount of chemical being sold. Amounts less than 10 kg year^{-1} used for research and similar purposes may be exempt. With less than 1 tonne year^{-1}, a simple base set is required comprising chemistry, production, uses, precautions; acute, skin and eye toxicity; mutagenicity *in vitro*; and simple ecotoxicity (degradation).

At 1 tonne year^{-1} (or 5 tonnes total) the base set plus a repeat-dose 28 day toxicity test, second mutagenicity test, screen for toxicity to reproduction, kinetics, and ecotoxicity are required. When the amount sold reaches 10 tonnes year^{-1} (50 tonnes total) the authority may require Level 1 tests, and when the amount reaches 100 tonnes year^{-1} (500 tonnes total) the authority shall require Level 1 tests. When the amount sold reaches 1000 tonnes year^{-1} (5000 tonnes total) the authority shall require Level 2 tests. Depending on the results obtained at all the different stages, the possibility exists for the authority to require tests to be carried out at different times from the above basic plan.

The Level 1 tests include a fertility study, a teratogenicity study, and sub-chronic toxicity, as well as further mutagenicity, kinetic, and ecotoxicity tests. The Level 2 tests include a further fertility (three generation) study if deemed appropriate as a consequence of the earlier tests, a teratology test in a second species, a developmental toxicity test (on peri-postnatal effects), as well as chronic toxicity, carcinogenicity, kinetics, and ecotoxicity tests.

Although full testing strategies are not yet agreed for all these effects it is hoped that these changes will substantially reduce hazards for men and women exposed to chemicals at work and in the environment.

10.7.4 Classification of Chemicals as Toxic for Reproduction

This classification has two major subdivisions.

10.7.4.1 Effects on Male or Female Fertility. This includes adverse effects on libido, sexual behaviour, any aspect of spermatogenesis or oogenesis, hormonal activity, or physiological response which would interfere with the capacity to fertilize, fertilization itself, or the development of the fertilized ovum up to and including implantation.

10.7.4.2 Developmental Toxicity. This is taken in its widest sense to include any effect interfering with normal development which is induced prenatally and may be manifest either pre- or postnatally. This includes embryofetal toxicity, death, abortion, retarded development, structural (teratogenic) effects, functional defects, and impaired postnatal mental and physical development up to and including normal pubertal development.

Chemicals are only classified as toxic to reproduction when they have specific, intrinsic, toxic potential to adversely affect reproduction. That is, the effects should not merely be secondary to other toxic effects. For example, it is well known that any treatment with high toxic doses of chemicals which cause severe inanition will reduce fertility or produce adverse effects on the developing fetus. Effects of this type would not lead to classification. This does not exclude substances that affect reproduction at around the same dose levels that cause other signs of toxicity, if it is thought that the effects on reproduction are not secondary to the other toxicity. Because of the above restrictions, information obtained from *in vitro* tests are only regarded as supportive data and would not normally be a basis for classification.

10.8 CATEGORIZATION

When classified, chemicals are also categorized. Allocation to Category 1 is made on the basis of human data, with evidence of reproductive toxicity to humans. Allocation to Category 2 is usually made on the basis of animal studies, with results suggesting that adverse effects would be likely with human exposure. For example, listing in Category 2 for fertility would require adverse results from work on one species, plus evidence of mechanism of action or site of action or structure–activity relationships to known active compounds, and evidence that the results would be likely to be relevant to humans. With less data, Category 3 or no classification would be appropriate. For listing in Category 2 on grounds of developmental toxicity, good animal data in one or more species, in the absence of marked maternal toxicity, and administered by a relevant route, would be expected. If the data are less convincing, or if there is the possibility of non-specific effects, or if only small changes in common variants are reported, or if there are only small changes in postnatal developmental tests, then it may be that Category 3 or no classification would be appropriate. Chemicals which are classified have to carry certain risk phrases on the labels such as 'May impair fertility' or 'May cause harm to the unborn child'.

10.9 LACTATION

When women are exposed to chemicals at work, there is concern not only with respect to pregnancy, but also for the lactation period. With improved creche facilities being provided by some employers, there is the possibility of women returning to work after the birth of a baby, but while the child is still being breast fed. There is also concern about contamination of breast milk from environmental pollution. In general, chemicals which have high fat solubility, and especially chemicals which are also poorly metabolized, may be expected to accumulate in breast milk. If such chemicals have toxic potential for the baby, they are labelled with a risk phrase 'May cause harm to breast fed babies'. Chemicals are not labelled in this way simply because they are present in milk. There must be harmful potential as well.

10.10 BIBLIOGRAPHY

S.M. Barlow and F.M. Sullivan, 'Reproductive Hazards of Industrial Chemicals'. Academic Press, London, 1982.

G.G. Briggs, R.K. Freeman, and S.J. Yaffe. 'Drugs in Pregnancy and Lactation'. Williams and Wilkins, Baltimore, MD, 3rd edn., 1990.

EEC, Council Directive of 5th June 1992 amending for the 7th time Directive 67/548/EEC, European Communities No. L 154/1, European Economic Community, Brussels, 1992.

FDA, 'Toxicological Principles for the Safety Assessment of Direct Food Additives and Color Additives used in Food'. Food and Drug Administration, Washington, DC, 1993.

O.G. Fitzhugh. 'Reproduction tests', in 'Modern Trends in Toxicology', eds. E. Boyland and R. Goulding. Butterworths, London, 1968, pp. 75–85.

'Guidelines for the testing of chemicals for toxicity (1982)', Report on Health and Social Subjects 27. HMSO, London, 1982.

International Conference on Harmonization, 'Harmonized Guideline for Detection of Toxicity to Reproduction for Medicinal Products', Federal Register Vol. 59, Food and Drug Administration, Washington, DC, 1994.

J.C. Lamb and P.M.D. Foster, eds. 'Physiology and Toxicology of Male Reproduction'. Academic Press, San Diego, CA, 1988.

D.R. Mattison ed. 'Reproductive Toxicology'. Alan R. Liss, New York, 1983.

H. Nau and W.J. Scott, eds. 'Pharmacokinetics in Teratogenesis', (2 volumes). CRC Press, Boca Raton, FL, 1987.

OECD, 'Guidelines for the Testing of Chemicals. Section 4: Health Effects, No. 414 Teratogenicity, No. 417 One Generation Reproduction Toxicity Study, No. 418 Two Generation Reproduction Toxicity Study'. Organization for Economic Co-operation and Development, Brussels, 1981, 1983, 1983 respectively.

J.L. Schardein. 'Chemically Induced Birth Defects'. Marcel Dekker, New York, 2nd edn, 1993.

A.R. Scialli. 'A Clinical Guide to Reproductive and Developmental Toxicology'. CRC Press, Boca Raton, FL, 1992.

F.M. Sullivan. Reproductive toxicity tests: Retrospect and prospect, *Hum. Toxicol.*, 1988, **7**, 423–427.

J.G. Wilson and Fraser F. Clarke eds. 'Handbook of Teratology', (4 volumes). Plenum, New York, 1977.

CHAPTER 11

Immunology and Immunotoxicology

JOHN H. DUFFUS

11.1 GENERAL CONCEPTS OF IMMUNOLOGY

Immunology is the science of the immune system: the natural mechanisms of the body responsible for its defences against invading micro-organisms and normal structures altered by the effect of micro-organisms or chemicals, or by natural processes including those that may lead to neoplasia.

The system comprises three interdependent sets of mechanisms: cellular immunity, humoral immunity, and non-specific immunity.

11.2 CELLULAR ('CELL-MEDIATED') IMMUNITY

Cellular ('cell-mediated') immunity is the system of specialized cells (phagocytic, antigen-presenting cells), by which an immunogen or, more commonly, an antigen (on an organism, cell, or protein; see below) is picked up and presented to T-lymphocytes (acting on surface receptors associated with MHC Class II histocompatibility antigens; see below), ultimately resulting in the development of an amplified clone (permanent line) of cells specialized for rapid recognition, further cell amplification, and reaction against that antigen. The latter process involves contact with the antigen by T-cells and the subsequent synthesis and release by them of specialized lytic factors, capable of destroying organisms and cells, and the release of inflammatory mediators. This process takes 24–48 h, hence the old term 'delayed-type' hypersensitivity applied to slowly appearing reactions, *e.g.* swelling of the skin at a challenged site.

Control of the reactive cell proliferation towards cell-mediated, or humoral immunity (see below), depends on the balance between the specific cytokines produced.

An antigen (see below) is any substance (or large molecule) able to evoke a specific humoral or cell-mediated (delayed-type) immune response. It does this because it presents specific paratopes (binding sites) recognizable by the immune system.

A hapten is a molecule about the size of an antigenic determinant (epitope), but itself too small to excite an immune response unless coupled to a carrier. It is able to combine with an antibody or the receptor of a sensitized cell.

Small molecules [*e.g.* certain metal ions, such as Ni(II) or Cr(III), or lower aliphatic aldehydes] may cause a specific immune response if they combine with a sufficiently large carrier molecule.

An antibody is a member of a superfamily of globulins (immunoglobulins) synthesized by cells of the immune system in response to exposure to an antigen, and with a tertiary structure adapted to combining with specific sites on that antigen (epitopes). After binding, many are capable of setting off the complement (see below) cascade and of attracting and activating macrophages and other types of leucocytes. The main types are immunoglobulin (Ig)G, IgM and IgE (all found in the bloodstream), IgA (secreted especially onto mucosal surfaces), and IgD. IgM and IgD occur as membrane-bound receptors on certain cells of the immune system. The formation and secretion of specific antibodies comprises a major part of the humoral immune response (see below).

The paratope is the specific site on an antibody or T-cell receptor that interacts by spatial complementarity with the epitope on the antigen. The closeness of the fit determines the specificity of the binding. It depends on the tertiary structure of the variable region of the antibody molecule, or on the corresponding structure of the T-cell receptor associated with MHC Class II molecules.

The idiotype is the specific antigenic determinant(s) on the variable domain of an immunoglobulin molecule or T-cell receptor characteristic of and defining its unique immune reactivity. By virtue of its need for appropriate spatial complementarity, it forms a 'mirror image' of the epitope on the relevant antigen.

Complement is a series of about 20 proteins, which form a cascading system of proteinases in the blood. Once triggered, there is very rapid (seconds) amplification of and production of the later stage factors, which are capable of lysing certain cells and micro-organisms, activating macrophages and other cells, causing adherence of micro-organisms to cells, and producing pharmacological effects on smooth muscle, endothelial, and other cells. Complement is responsible for initiating some of the processes of acute inflammation. The classical cascade is normally triggered by conformational changes in IgG or IgM molecules after combination with their specific antigen (paratope), and the alternative cascade by certain microbial polysaccharides and other non-immune specific activators.

Histocompatibility antigens are complex multigene, multiallele superfamilies of antigens on cell surfaces that give them immunological specificity. They determine whether a cell will be recognized as 'self' or 'foreign'. Antigens have to be presented in association with certain histocompatibility antigens if they are to excite an immune response. The surface expression of these antigens depends on the activity of the cells. There are major (MHC Class I) and minor (MHC Class II), but still important, histocompatibility antigens.

Antigen-presenting cells are specialized dendritic cells and macrophages, that take up antigens and present processed peptide fragments to cell-surface receptors on T- and some B-cells.

T-cells are specialized types of lymphocytes produced in the bone marrow and matured in the thymus (hence 'T-cell'). They are able to induce, regulate, and produce specific immune reactions in response to an antigen on antigen-presenting cells associated with MHC molecules.

Mature ('primed' or 'armed') T-cells are directly responsible for cell-mediated immunity, and they form an essential early stage in most B-cell (see below) responses. Some T-cells are very long lived and so form part of immunological memory.

The early phase of the immune response may be directed by the activities of specialized T-cells towards magnified cell-mediated and humoral responses (T_{H1} or T-helper cells) or towards relative inhibition of the immune response (T_s or T-suppressor cells; also called TH_2). The latter are probably very important in controlling autoimmunity. The T_{H1} and TH_2 cells probably act by secreting specific types of cytokines.

B-cells are the system of lymphocytes and plasma cells that secrete antibodies on exposure to appropriate antigens, either by direct stimulation, or more commonly *via* a T-cell-dependent step. B-cell precursors develop in the bone marrow and are later seeded into peripheral lymphoid tissues. The name 'B-cells' comes from their initial recognition in the chicken, where they mature in a specialized organ, the Bursa of Fabricius.

GALT, BALT, MALT are abbreviations applied to gut (G), bronchus (B), and mucosa (M) associated lymphoid tissue (ALT) and are specialized collections of lymphoid cells associated with the corresponding anatomical structures. They respond to antigens presented at those sites as part of the protective response against micro-organisms.

11.2.1 Lymphocyte Mitogens

A lymphocyte mitogen is a substance capable of reacting with surface receptors on lymphocytes, causing proliferation.

There are specific T- (concanavalin A) and B-cell (pokeweed mitogen) mitogens, and more general panmitogens (phytohaemagglutin). The mitotic responses of cells to a mitogen gives some impression of the state of the relevant subpopulation.

11.3 HUMORAL IMMUNITY

Humoral immunity is the process by which a specific antibody is produced in response to a specific antigen. It depends on antigen uptake and presentation to certain types of T-cells which, in turn, and by mechanisms largely dependent on specific cytokines, cause the proliferation of lymphocytes and plasma cells that secrete specific antibodies against that antigen.

Many antigens first excite IgM formation, followed by IgG, but some lead especially to IgE antibody, and others, such as micro-organisms at mucosal surfaces, result in the local secretion of specific IgA.

The combination of an antigen with its unique antibody results almost immediately in the release of inflammatory mediators, including activation of the complement cascade, and excitation of the acute inflammatory response. The resultant reactions are apparent within a few minutes to about 1 hour.

11.4   NON-SPECIFIC ('INNATE') IMMUNITY

Non-specific ('innate') immunity is the most primitive type of immune system, comprising phagocytes and complement, lysozyme, and interferon which react against foreign antigens without the specificity of the T- and B-cell-dependent mechanisms.

11.5   ALLERGY

Allergy is a state of extreme sensitivity (hypersensitivity; see below) due to prior exposure of the body to an antigen, to which it has become highly responsive. On subsequent exposure, an excessive or otherwise abnormal immune response occurs, usually very rapidly, resulting in a disproportionately severe reaction, *e.g.* the asthmatic response to an inhaled antigen (due in part to combination of the antigen with specific IgE antibody on mast cells, and the local release of potent pharmacological mediators), or the dermatitis caused by exposure of the skin of a sensitized subject to nickel, *etc.*

An allergen is a substance or material capable of exciting an excessive immune response, sometimes due to the abnormal production of IgE antibody or to the appearance of sensitized T-cells (lymphocytes).

11.6   SENSITIZATION

Sensitization is a state of increased immunological reactivity, due to prior exposure to an antigen. Subsequent exposures lead to an amplified cellular and/ or humoral immune response. Sensitization is commonly but confusingly used to refer to an excessive or otherwise inappropriate reaction, which would better be termed 'hypersensitivity' (see below).

The respiratory tract and skin are the two main routes of sensitization but there is increasing concern about possible reactions to food additives.

11.7   HYPERSENSITIVITY

Hypersensitivity is a condition in which prior exposure to an antigen has led to immunological sensitization following which subsequent exposures result in an inappropriate type and excessive immune (allergic) response, *e.g.* allergic contact dermatitis, asthma, and anaphylaxis (see Type I–IV Hypersensitivity Responses below).

A convenient method of classifying hypersensitivity responses is in terms of their pathogenic mechanism, although in practice more than one process may be active at a time. The conventional classification is as follows:

Type I: Mast cells bind specific IgE *via* their surface F_c receptors. Exposure to the specific antigen results in degranulation and release of mediators.

Type II: Antibody combines with antigens on cells, leading to complement activation and cell lysis or to cell destruction by Natural Killer cells (sometimes known as 'Antibody Dependent Cell Toxicity').

Type III: Preformed complexes of antibody and antigen are deposited in tissues, leading to local activation of complement and tissue damage due to a localized inflammatory response.

Type IV: Sensitised T-cells are activated by contact with the specific antigen, resulting in the synthesis and release of specific cytokines, the recruitment and activation of macrophages, and a local inflammatory response.

11.8 TOLERANCE

Tolerance is a state in which the body does not react against foreign antigens. It may be caused by exposure to the antigens in fetal life, or it may be produced by particular patterns of exposure to soluble antigens.

11.9 VACCINES

Vaccines are live or dead organisms, or purified antigens, often combined with an adjuvant (see below), used to stimulate immunity against a pathogenic organism.

11.10 AUTOIMMUNITY

Every protein and many other components in the body contain many epitopes. The body does not normally react immunologically against itself because of control by a complex network of cells, cytokines, and naturally occurring antibodies. This may break down in certain infections, after chemicals and even mechanical trauma, and in later life, when autoreactive antibodies and T-cells may appear, directed against 'self components'. This is the phenomenon known as 'autoimmunity'. The result may be disease, of which the best known in man include Hashimoto's thyroiditis, pernicious anemia, and rheumatoid arthritis ('collagen disease').

11.11 GRAFT *VERSUS* HOST DISEASE

After transplantation of immunocompetent lymphocytes, the graft may produce a serious and even fatal immune reaction against antigens in the host. Allogeneic and syngeneic are terms referring to the histocompatibility (transplantation) antigens of individuals of the same species.

Syngeneic animals have identical antigens and freely accept tissue grafts from each other. Allogeneic animals differ in one or more antigens and react against

cross-grafts unless immuno-suppressed or made tolerant by immunopharmacolo-
gical treatment.

11.12   IMMUNODEPRESSION (IMMUNOSUPPRESSION)

Immunodepression (suppression) is an impairment of the immune response,
which may affect any or all of the cellular, humoral, or innate systems. Many
substances and disease states may cause a degree of immunodepression.

Therapeutic immunosuppression is commonly produced by treatment with
corticosteroids or cytotoxic agents, as well as specific immunosuppressants, such
as cyclosporin A.

11.13   IMMUNOENHANCEMENT

Immunoenhancement is an increase in the activity of one or more components of
the immune system. It may be produced by such chemicals as levamisole or
thymopentin, and by carefully timed doses of certain cytotoxic agents, notably
cyclophosphamide, which probably act by destroying suppressor T-cells.

11.14   ADJUVANTS

Adjuvants are more or less complex substances that greatly enhance the immune
response, and which may cause it to be of a limited type, *e.g.* produce almost
exclusively IgE or G antibodies, sensitized cells *etc.*

Freunds Complete Adjuvant, a very powerful adjuvant, is a mixture of mineral
hydrocarbons and killed *Mycobacterium tuberculosis*. Conventional vaccines used in
man rely on an aluminium hydroxide or phosphate adjuvant.

11.15   IMMUNOTOXICITY

Immunotoxicology is a general term denoting a toxic effect on or exerted by the
immune system. Often it is restricted to actions on the immune system and their
immediate consequences. At least four types of unwanted consequences may
result from effects of chemicals on the immune system:

1. Immune depression: alterations of host defence mechanisms making them
   less effective.
2. Immune enhancement: stimulation of normal immune responses to
   harmful levels.
3. Allergy or hypersensitivity: qualitatively abnormal immune responses.
4. Autoimmunity: immunological attack on self tissues.

It is of fairly recent interest to toxicologists, although it had been shown as
early as 1896 that ethanol intoxication decreased the natural resistance of rabbits
to experimental streptococcus infection. Modern concern about immunodepression stems largely from the observation that polychlorinated biphenyls caused

signs of immune deficiency in experimental animals with measurable decreases in immunoglobulin concentrations, especially of IgM and IgA. This was associated with persistent skin infections and bronchitis-like respiratory symptoms. Concern was reinforced by the observation that Michigan residents who had been exposed to polybrominated biphenyls showed immune dysfunction.

Generally, toxicological damage to a target organ is directly related to the concentration of the toxic agent or its metabolite in the organ affected, but some immune-mediated effects do not conform in a simple way to this concept because they result from subsequent physiological consequences. Most obviously, depression of the immune system may result in death from infection with a virulent organism. As another example, compounds of mercury and gold can damage the kidneys as a secondary consequence of their interaction with the immune system as a result of the deposition in these organs of immune complexes.

11.15.1   Health Consequences

Health consequences include the following:

1. Decreased resistance to infection. This is commonly seen in cancer patients receiving long-term chemotherapy.
2. Increased incidence of cancer. Cancer patients receiving long-term chemotherapy show a higher incidence of malignancies.
3. Autoimmune disorders. A number of postulated drug-induced autoimmune diseases such as lupus syndrome, myasthenia gravis, and haemolytic anemia have been described.
4. Hypersensitivity reactions. Hypersensitivity reactions have been noted following the use of immunomodulating agents, and febrile reactions with hyperpyrexia, chills, malaise, and asthenia being frequent complications which follow.
5. Alteration of drug metabolism. Inhibition of hepatic drug metabolism is a common consequence of pharmacological immunostimulation and has been reported with most immunomodulating agents.

11.16   BIBLIOGRAPHY

A.D. Dayan, R.F. Hertel, E. Heseltine, G. Kazantzis, and E.M.B. Smith, eds. 'Immunotoxicity of Metals and Immunotoxicology'. Plenum, New York, 1990.

J. Descotes, 'Immunotoxicology of Drugs and Chemicals'. Elsevier, Amsterdam, 2nd edn, 1989.

P.J. Lachmann, D.K. Peters, F.S. Rosen, and M.J. Walport, eds. 'Clinical Aspects of Immunology'. Blackwell Scientific, Boston, MA, vols. I–III, 1993.

K. Miller, J.L. Turk, and S. Nicklin, 'Principles and Practice of Immunotoxicology'. Blackwell, Oxford, 1992.

D.S. Newcombe, N.R. Rose, and J.C. Bloom, eds. 'Clinical Immunotoxicity'. Raven Press, New York, 1992.

W.E. Paul, ed. 'Fundamental Immunology'. Raven Press, New York, 2nd edn, 1989.

I. Roitt, 'Essential Immunology'. Blackwell Scientific, London, 8th edn, 1994.

I. Roitt, J. Brostoff, and D. Male, eds. 'Immunology'. Mosby, London, 4th edn, 1996.

WHO, International Programme on Chemical Safety, 'Principles and Methods for Assessing Direct Immunotoxicity Associated with Exposure to Chemicals', World Health Organisation, Geneva, in the press.

CHAPTER 12

Skin Toxicity

R.D. ALDRIDGE

12.1 INTRODUCTION

The skin is the largest organ of the body. It interfaces with the environment, both maintaining the equilibrium of the internal milieu and preventing the ingress of environmental toxins. The structure consists of an inner layer called the dermis which sits astride the subcutaneous fat and adjoins the overlying epidermis. The cellular content of the dermis comprises predominantly fibroblasts and mast cells. The fibroblasts produce the collagen and elastin fibres which give the skin its tensile strength. These fibres are embedded in a ground substance of sugar-containing molecules known as glycosaminoglycans which are also produced by the fibroblasts. This extracellular matrix situated at the dermo–epidermal junction is now known to have a significant effect on the growth of cells of the epidermis. The dermis also contains cells of the monocyte/macrophage system which are part of the body's cellular defence mechanism and believed to be involved with both allergic and irritant dermatitis responses.

The epidermis covers the dermis and contains a basal layer of replicating cells. Above the basal layer several layers of terminally differentiating cells evolve into the stratum corneum. The latter is a water impenetrable, anucleated, cell layer that interfaces with the environment. This extraordinary barrier is highly resistant to mechanical and chemical assault and an understanding of its structure and development is essential to an understanding of the toxicology of the skin.

The stratum corneum is nowadays thought of as a two-compartment system consisting of lipid depleted protein-rich corneocytes bound with complex bilayer lipids. The formation of this protective structure occurs predominantly in the stratum granulosum which is situated just under the stratum corneum.

Within the stratum granulosum, enzymes transform the lipid content of the cell into the complex bilayers which are then excreted, the remaining protein-rich cell being termed a corneocyte. The control of stratum corneum production is intimately associated with perturbation of its barrier function. Thus damage to the stratum corneum will generate both enzyme activity within the stratum granulosum and increased replication of the basal keratinocytes.

12.2   PERMEATION

Penetration of the stratum corneum is the rate-limiting step in percutaneous absorption. If the barrier is damaged by vesiculants such as acids or alkalis, absorption will be enhanced. Permeation of toxins through the intact stratum corneum is dependent upon the lipid solubility of the toxin, the concentration of the toxin at the skin surface, and the nature of the carrier vehicle within which the toxin exists. Penetration will be influenced by the thickness of the stratum corneum and the presence of skin appendages such as hair follicles and sweat glands. The study of regional variations in absorption has been limited but it is recognized that considerable differences exist, with permeation in the periorbital or genital skin being rapid, and in the hyperkeratotic skin of the palms and soles extremely slow.

The cutaneous blood flow primarily serves the purpose of thermoregulation and vastly exceeds the metabolic requirements of the skin. It is generally accepted that this excess blood flow results in rapid chemical clearance from the subcutis, preventing any build-up or reservoir effect that would inhibit penetration. Under these circumstances Fick's law of diffusion applies, *i.e.* the flux (J) is proportional to the concentration of the penetrant (C_v) thus:

$$J = K_p C_v$$

where K_p is the permeability constant.

The permeability constant enables the absorption to be calculated from various concentrations of any given penetrant. A variety of *in vitro* techniques have been developed for the calculation of absorption. Care must be taken in extrapolating the results of such *in vitro* studies to the dynamic *in vivo* situation. Several *in vitro* systems do produce sufficiently consistent results which, in conjunction with such *in vivo* studies as are practicable, enable useful predictions to be made on dermal absorption of certain toxins.

12.3   DERMAL TOXICOLOGY

Dermal toxicology must therefore consider both the direct effect of environmental toxins on the skin and the systemic effects of cutaneous absorption. Substances that are absorbed through other routes may have a particular affinity for the skin *e.g.* itraconazole which preferentially accumulates in keratin. In these circumstances the skin may be the primary target of damage from toxins that are not absorbed cutaneously. Conversely many dermally absorbed toxins may have little direct effect on cutaneous function, but possess profound systemic effects. Thus, the plasticizer triorthocresyl phosphate is a neurotoxin with little adverse cutaneous effect. For other toxins, the cutaneous and systemic effects differ. Thus PVC is a narcotic in high dosage and can cause the rare hepatic haemoangiosarcoma: in the skin digital scleroderma may develop. In humans dioxin is remarkably non-toxic in comparison with many other species but it can cause deranged hepatic function when absorbed in large quantities. On the skin it can produce a characteristic abnormality known as chloracne. Toxins freely passing through the dermal barrier are generally lipophilic, a characteristic that favours neurotoxicity. Irrespective of any systemic effect, the commonest cutaneous response elicited by environmental toxins is dermatitis.

12.4 DERMATITIS

Dermatitis is an inflammatory condition which can be caused by irritants or allergens. If the latter, damage is mediated through macrophages and sensitized lymphocytes following recognition of antigen in contact with the skin. In irritant contact dermatitis the irritant itself inflicts damage which is compounded by the body's defence mechanism. Both these disorders give rise to a systemic response which may have effects elsewhere. Thus, in the case of allergic contact dermatitis, secondary eczematization is often seen and in irritation the phenomena of conditioned hyperirritability can effect responses to irritation outside the primary site of contact. Consideration of eczema is outside the scope of this review but many toxins can have irritant or allergic potential and, if dermatitis arises, increased absorption will occur through the inflamed skin.

12.5 TOXIN ACCUMULATION AND METABOLISM

Toxins may accumulate in the skin as a result of absorption or by dermal localization after ingestion, inhalation, or injection. Elsewhere in this book metabolism by the liver of ingested toxins has been described in detail. Similar detoxifying enzymes exist within the skin, which has been estimated to have between 2 and 6% of the detoxifying potential of the liver. While enzyme activity has been identified in epidermis, epidermal appendages, and dermis, the major site of detoxification is the epidermis. Most xenobiotics absorbed through skin are lipophilic, and Phase 1 metabolism involves the introduction of a polar group by hydroxylation of a carbon, nitrogen, or sulfur atom using molecular oxygen. As in the liver, this is accomplished by a mixed-function oxidase, cytochrome P450 using molecular oxygen with the formation of water as a by-product. These enzymes are the terminal component of an electron transport system that is found in the microsome by which NADPH (reduced nicotinamide–adenine dinucleotide phosphate) transfers high-potential electrons to a flavoprotein which then conveys them to adrenodoxin, a non-haeme iron protein. Adrenodoxin then transfers an electron to the oxidized form of cytochrome P450. The reduced form of P450 then activates O_2.

It is clear that in the liver P450 is not a single entity and a variety of individual forms exist that exhibit differing substrate specificity. Investigation of these enzyme systems in the skin has been hampered by the structural resistance of mature skin to the homogenization procedures used in hepatic investigations. Many of the skin studies have been undertaken with neonatal rat skin but do suggest a variation of P450 enzymes similar to that found in the liver, and furthermore these can be induced by the application of topical toxins. The enhancement appears to result from both increased gene expression and enzyme induction. Phase 2 metabolism involves a conjugation of the Phase 1 reactant to form water-soluble products which are more easily excreted. Skin possesses Phase 2 enzymes such as phenol uridine 5′-diphosphate (UDP)-glucuronosyltransferase and glutathione S-transferase, and thus the complete metabolism of a toxin to a water-soluble excretable form can occur within the epidermis.

In the process of metabolizing xenobiotics, highly reactive intermediate products can be formed from substances which are non-toxic or even inert. The effect of these toxic metabolites varies. Thus the commonly used topical anti-acne preparation, benzoyl peroxide, does not become effective until it is metabolized. In addition to its irritant and proliferative effects, the metabolized benzoyl peroxide can act as an allergen. Most common contact allergens are easily absorbed, low molecular mass substances which become allergic by haptenization with skin constituents. The resulting product is recognized by the immune system as foreign and allergenic. It should therefore occasion little surprise that several substances, *e.g.* β-phenylenediamine, are known to become allergic only after metabolism to active metabolites by cytochrome P450.

In addition to irritation and allergy, toxic metabolites produced by skin metabolism are capable of inducing carcinogenesis. Carcinogenesis is believed to be a multistage process involving initiation, promotion, and progression. Environmental toxins and other damaging insults are being increasingly incriminated in the process of carcinogenesis. Indeed, it is difficult not to accept an association between the life-long exposure of the skin to environmental assault from sun and chemical carcinogens and it being the commonest site of human malignancy.

It seems unlikely that any single environmental agent is responsible for all stages of tumour development. Initiation probably involves covalent binding by the toxic molecule to DNA (deoxyribonucleic acid). Subsequent tumour development is dependent upon epidermal hyperplasia, and many irritants and mitogens capable of inducing epidermal hyperplasia are recognized tumour promotors. The suggestion by Potts in 1775 that scrotal cancer in sweeps originated from their occupational exposure to soot initiated a series of investigations which culminated in the identification of the polycyclic aromatic hydrocarbons (PAHs) as the factors responsible for the carcinogenic potential in soot, oil, and coal tars. The PAHs are metabolized by cytochrome P450-associated aryl hydrocarbon hydroxylase. It is believed that environmental PAHs derived from incomplete combustion of fossil fuel may be among the commonest chemical carcinogens. The concept, now generally accepted, that tumours are initiated by somatic mutation within the cell arose from the finding that the binding activity of polycyclic hydrocarbons to epidermal DNA correlated with the development of tumours. The mechanism by which binding to the DNA induces carcinogenesis has not been fully elucidated, but one component may involve mutation of proto-oncogenes as *ras* oncogene has been detected in some basal cell and squamous cell carcinomata.

12.6   APOPTOSIS

Physiological cell death or apoptosis is a major regulatory mechanism controlling both normal and abnormal cell growth. It is a non-inflammatory, genetically controlled, energy-dependent process which results in the elimination of unnecessarily damaged or harmful cells.

Apoptosis can be induced or suspended by many environmental factors such as UV light or cytotoxic drugs. During apoptosis there is variation in the expression of pro-oncogenes and anti-oncogenes implicating their involvement in programmed cell death. It also seems likely that the human cancer suppressor gene p53 exerts some of its effect by inducing apoptosis. The dermal intracellular level of p53 is increased in response to DNA-damaging doses of UV-B (315–280 nm). Apoptosis has been identified in both squamous cell carcinomas and basal cell carcinomas, both tumours in which UV-B has a major aetiological role. UV-B can, however, also induce p53 gene mutation, which renders it ineffective and it may be the consequence of loss of p53 control that leads to tumour proliferation.

The induction of apoptosis by dermal toxins has not been studied in any depth but it seems probable that if a toxic assault results in irreparable DNA damage, apoptosis would be induced. The relationship of apoptosis to terminal differentiation such as occurs in the development of the stratum corneum from the basal keratinocyte, has yet to be determined, but if indeed epidermal terminal differentiation is a specialized form of apoptosis, toxic interference with the process may be another mechanism that underlies carcinogenesis.

12.7   OTHER METABOLITES

A number of other enzymes of metabolic importance are known to exist within the epidermis. The most well characterized are those responsible for dermal steroid metabolism. The activity of these enzymes differs in body sites and between persons. Thus 5α reductase, the enzyme responsible for mediating the reduction of testosterone to the more active dihydrotesterone is believed to vary in concentration at various body sites, and this variation in distribution may be of importance in development of secondary sexual characteristics. In a similar manner, aromase enzymes, which have been found to be present at the outer root sheaf of the hair follicle, are present at 5 times greater concentration in women than in men. Since aromase mediates the conversion of testosterone and androstenedione to estradiol and estrogen, this enzyme may well be responsible for the differing pattern of hair loss in men and women. It is as yet unknown whether these enzymes play any part in the metabolism of exogenous steroids or other xenobiotics.

Many other enzyme systems exist in the skin that are involved in the metabolism of endogenous substrates. In some cases the absence of these may give rise to skin abnormalities, *e.g.* deficiency of cholesterol sulfate in the epidermis of those with x-linked ichthyosis. Such deficiencies may, by the alteration of the cutaneous barrier, have an effect on the fate of applied toxins, but direct participation in metabolism has not been demonstrated.

12.7.1 Peroxisomes

Peroxisomes (microbodies) are, like mitochondria, a major site of oxygen utilization. It is believed that they are present in all cells, and recently peroxisomes have been induced in cultured keratinocytes. Catalase comprises some 40% of total protein content of peroxisomes and it seems likely that these organelles are the source of the catalase activity demonstrated in human skin.

Peroxisomes are diverse organelles and in addition to catalase contain differing sets of enzymes that can adapt to changing conditions. Unfortunately, little is presently known of the enzyme system existing within the dermal peroxisomes but in general the peroxidase enzymes use molecular oxygen to remove hydrogen ions from organic substrates. This oxidation process produces hydrogen peroxide (H_2O_2) and catalase utilizes the hydrogen peroxide to oxidize such substances as phenols and alcohol with the production of H_2O. It seems likely that this system may participate in the metabolism of phenol and other toxins applied to the skin. Such metabolism is likely to generate free radicals. Free radicals are also generated by the action of UV-B light on the skin. Indeed, it is likely that many of the adverse effects of light on the skin are mediated by free radical production. The effect of light in the UV-A spectrum (400–315 nm) can be enhanced by a number of plant derivatives, *i.e* fluorcoumarines, of which psoralens, used both topically and systematically in PUVA therapy, are the most well characterized.

The excitation of a ground state molecule by energy transfer from another species is termed 'sensitization' and the deactivation of an excited species 'quenching'. Both phenomena are of central importance in organic photochemistry. Many chemicals are capable of producing dermal phototoxicity and the potential effects of UV light on absorbed xenobiotics are immense and, as yet, poorly understood.

Reactive oxidants can therefore be generated in the skin not only from normal endogenous processes but also from exogenous sources by redox cycling or electromagnetic radiation. Toxicity of reactive oxidants is felt to be due to non-specific attack upon structural cell constituents. These effects may vary from destruction to physiological disturbance of growth including apoptosis and cell signalling. Oxidative stress is mitigated by scavenging systems such as catalase and superoxide dismutase. If these protective measures fail and damage occurs, repair is effected by a mixture of enzyme and non-enzyme processes. It is outside the scope of this review to discuss the complex antioxidant system in depth, but our knowledge of the system is incomplete, as is our knowledge of the susceptibility of the components of the system to damage from cutaneously applied toxins.

12.8   CONCLUSIONS

In this introductory chapter an attempt has been made to outline the physiology and structure of the skin barrier. An outline has been given of the metabolism of absorbed xenobiotics and the manner in which the skin deals with other chemical and UV assault.

It is apparent that our understanding of skin toxicity is as yet elementary but the development of new molecular biological techniques and the increasing success of keratinocyte culture may well enable us to circumvent the difficulties inherent in the study of an organ whose physical structure resists homogenization. It is to be hoped that the processes underlying the essential protective function of skin will in the near future become increasingly better understood.

12.9 BIBLIOGRAPHY

M. Aumailley, and T. Kreig, Structure and function of the cutaneous extracellular matrix, *Eur. J. Dermatol.* 1994, **4**, 271–280.

O. Baadsgaard, and T. Wang, Immune regulations in allergic and irritant skin reactions, *Int. J. Dermatol.* 1991, **30**, 161–170.

D.R. Bickers, T. Dutta-Choudbury, and M. Mukhtar, Epidermis: A site of drug metabolism in neonatal rat skin. Studies on cytochrome P450 content and mixed function oxidase and epoxide hydrolase activity, *Mol. Pharmacol.*, 1982, **21**, 239–247.

P. Brookes and P.D. Lawley, Evidence for the binding of polynuclear aromatic hydrocarbons to the nucleic acids of mouse skin: Relation between carcinogenic power of hydrocarbons and their binding to deoxyribonucleic acid, *Nature*, 1964, **202**, 781–784.

J.W. Cook, I. Hieger, E.L. Kennaway, and W.V. Mayneord, The production of cancer by pure hydrocarbon. Part 1, *Proc. Roy. Soc. London*, 1932, **111**, 455–484.

J.M. Coxon and B. Halton, 'Organic Photochemistry'. Cambridge University Press, Cambridge, UK, 1987.

P.H. Dugard, 'Skin permeability theory in relation to measurements of percutaneous absorption in toxicology', in 'Dermatotoxicology', eds. F.N. Marzulli, and H.I. Maibach. Hemisphere, Washington, DC, 3rd edn, 1987, pp. 95–120.

P.M. Elias and C.K. Menom, Structural and lipid biochemical correlates of the epidermal permeability barrier, *Adv. Lipid Res.*, 1991, **24**, 1–26.

European Centre for Ecotoxicology and Toxicology of Chemicals, 'Monograph 20. Percutaneous Absorption'. European Centre for Ecotoxicity and Toxicology of Chemicals, Brussels, 1995.

A.R. Haake and R.R. Polakowska, Cell death by apoptosis in epidermal biology, *J. Invest. Dermatol.*, 1993, **101**, 107–112.

B.N. Kemppainen and W.G. Reifenrath, eds. 'Methods of Skin Absorption'. CRC Press, Boca Raton, FL, 1990.

W. Levin, Functional diversity of hepatic cytochrome P450, *Drug Metab. Dispos.*, 1990, **18**, 824–830.

G.K. Menon, D. Placzek, M. Hincenbergs, and M.L. Williams, Lovastatin induces peroxisomes in cultured keratinocytes, *J. Invest. Dermatol.*, 1989, **92**, 480.

H. Mukhtar and W.A. Khan, Cutaneous cytochrome P450, *Drug Metab. Rev.* 1989, **20**, 657–673.

H. Mukhtar, H.F. Merck, and M. Athar, Skin chemical carcinogenesis, *Clin. Dermatol.* 1989, **7**, 1–10.

B.J. Nickoloff and Y. Naidu, Perturbation of epidermal barrier function correlates with mutation of cytokine cascade in human skin, *J. Am. Dermatol.*, 1994, **30**, 535–546.

R. Pohl, R.M. Philpot, and J.R. Fouts, Cytochrome P450 content and mixed function oxidase activity in microscomes isolated from mouse skin, *Drug Metab. Dispos.*, 1976, **4**, 442–450.

M.E. Sawaya, Clinical approaches to androgenetic alopecia in women, *Curr. Opin. Dermatol.*, 1993, **11**, 91–95.

E.E. Sisskin, T. Gray, and J.C. Barret, Correlation between sensitivity to tumour promotion and sustained epidermal hyperplasia of mice and rats treated with 12-O-tetradecanoylphorbol 13-acetate, *Carcinogens*, 1982, **3**, 403–407.

R.M. Tyrrell and M. Pidou, Singlet oxygen involvement in the inactivation of human fibroblasts by UVA (334 nm and 365 nm) and near visible (405 nm) radiations, *Photochem. Photobiol.*, 1989, **49**, 407–412.

J.G. Van der Schroeff, C.M. Evans, A.J.M. Boot, and J.L. Bos, *ras* oncogenes mutation in basal cell carcinomas and squamous cell carcinomas of human skin, *J. Invest. Dermatol.*, 1990, **94**, 423–425.

F. Vecchini, K. Mace, J. Magdalou, Y. Mahe, B.A. Bernard, and B. Shroot, Constitutive and inducible expression of drug metabolizing enzymes in cultured human keratinocytes, *Br. J. Dermatol.*, 1995, **132**, 14–21.

K. Weber-Matthieson and W. Sterry, Organization of the monocyte/macrophage system of normal human skin, *J. Invest. Dermatol.*, 1990, **95**, 83–90.

CHAPTER 13

Respiratory Toxicology

RAYMOND AGIUS

As mentioned in Chapter 1, the respiratory tract is a very important organ of first contact for most environmental exposures. An adequate understanding of the toxicology of the respiratory tract requires some basic knowledge of the anatomy of this system and an outline is described below.

13.1 RELEVANT FUNDAMENTALS OF RESPIRATORY STRUCTURE AND FUNCTION

The nose is the organ in the respiratory tract that is first exposed to inhaled agents. It is provided with a convoluted surface over the turbinates or conchae. Its main channel is to the nasopharynx and hence larynx (voice box) but it also drains tears from the eyes and communicates with the sinuses.

The airways of the lungs derive from the trachea (wind pipe), which is a continuation of the larynx, by progressive division into two (or more) branches. Those airways beyond the trachea that contain cartilage are called bronchi. The airways lacking in cartilage beyond the bronchi are the bronchioles. These lead into hollow spaces called alveoli which have a diameter of about 0.1 mm each. There are approximately 300 million alveoli and their total surface area is about 140 m^2. The conducting airways are lined by cells with cilia (small motile surface projections). Interspersed between these cells are mucus-secreting cells.

Secreted mucus spreads over the cilia which direct it upwards to the larger airways by rhythmic wave-like movements, thus helping to clear deposited dusts. The respiratory units (*i.e.* the alveoli and the smallest bronchioles called respiratory bronchioles) are responsible for the exchange of gases. They are lined mainly by flat, extremely thin cells which permit easy diffusion of oxygen through them from the air in the alveolar spaces to the blood in the capillaries, and easier diffusion of carbon dioxide in the opposite direction. Alveolar macrophages are very abundant, which are mobile phagocytic cells mainly responsible, among other functions, for the ingestion of foreign matter. The smooth lining of the outside of the lungs and the inside of the chest wall is called the pleura.

13.2   DEPOSITION AND HOST DEFENCE OF INHALED DUSTS AND MISTS

Aerosol is an all-embracing term including all airborne particles small enough to float in the air. Dusts are solid particles dispersed in air. Mists are liquid droplets formed by the condensation of vapours (usually around appropriate nuclei), or the 'atomization' of liquids. The aerodynamic diameter of a particle is the diameter of a sphere of unit density that would settle at the same rate as the particle in question. When airborne particles come into contact with the wall of a conducting airway or a respiratory unit they do not become airborne again. This constitutes deposition and can be achieved in one of the following four ways:

- Sedimentation. Sedimentation is settlement by gravity and tends to occur in the larger airways from the nose downwards.
- Inertial impaction. This occurs when an airstream changes direction especially in the nose but also in other large airways.
- Interception. Interception applies mainly to irregular particles such as asbestos or other fibrous dusts which by virtue of their shape can avoid sedimentation and inertial impaction. However, they are intercepted by collision with walls of bronchioles especially at bifurcations or if the fibres are curved.
- Diffusion. Diffusion is the behaviour of very small aerosol particles which are randomly bombarded by the molecules of air through Brownian movement. It significantly influences deposition beyond the terminal bronchioles.

Most particles larger than 7 µm in aerodynamic diameter are deposited in the nose or throat during breathing at rest. However there is a wide variation in the efficiency of this among apparently normal subjects. Moreover, conditions that favour mouth breathing, *e.g.* high ventilation rates and obstructive disease of the nasal airways, will cause large particles to bypass this filter. Alveolar deposition is appreciable at particle aerodynamic diameters of between 1 and 7 µm (respirable particles) with the smaller particles being more likely to be deposited. There is evidence that very fine particles of less than say 0.1 µm can be particularly harmful. During exertion, an increase in tidal volume (*i.e.* the volume of air inspired with each breath) and particularly in respiratory minute volume (*i.e.* the product of tidal volume and the number of breaths per minute) is the single most important determinant of the total load of particles in the alveoli and hence the total volume of particles deposited for a given aerosol.

Several other factors may influence particle deposition. Insoluble particles deposited in the conducting airways are propelled towards the larger airways by the cilia and then rapidly coughed or swallowed. This may be delayed by factors such as tobacco smoking. In the respiratory units, ingestion by macrophages is necessary before the particles are carried to the larger airways. Particles may also penetrate the deeper lung tissue where they may stay for years or be transported by macrophages to the lymph nodes.

Thus, in a sense, the respiratory tract tends to behave like a very complex elutriator with the particles of larger aerodynamic diameter depositing in the nose and larger airways, while smaller particles deposit in the smaller airways and alveoli; a proportion of the very smallest particles are not deposited at all. A similar analogy may be extended to the handling of gases and vapours by the lung. (Vapours are substances in the gaseous phase at a temperature below their boiling point.) Gases that are very water-soluble such as ammonia will partition readily into the aqueous phase lining the mucous membranes and thus will achieve higher concentrations around the cells of the eyes and nose. Other less water-soluble gases or highly lipid-soluble vapours tend to be absorbed by the respiratory units of the lung.

13.3 DISEASES MAINLY OF THE AIRWAYS

13.3.1 Irritant Effects of Gases

Sulfur dioxide, nitrogen dioxide, ozone, ammonia, and chlorine are examples of such gases. These gases produce their harmful effects by irritating eyes, airways, and even the respiratory units of the lungs. Many of them may be detected by their smell and irritant effect, but if evasive action is not taken in time, and if exposure is high enough, they can produce severe damage throughout the lungs.

Exposure to ammonia and chlorine may occur as a result of industrial accidents. High levels of nitrogen dioxide can be encountered in agriculture (silo filling), during arc welding, as a result of shot firing in the mines, and in the chemical industry. It can achieve high levels in the vicinity of internal combustion exhausts. Ozone is usually a secondary pollutant. Sulfur dioxide is produced by the combustion of sulfur-containing fossil fuels.

Sulfur dioxide, chlorine, and ammonia are highly irritant and cause pain in the eyes, mouth, and chest. In high concentrations they can produce inflammation of the lining of the lungs and this causes breathlessness and may be fatal. (See chronic effects below.)

Nitrogen dioxide has less effect on the eyes, nose, and mouth but can cause severe inflammation of the lungs. It is important to realize that although symptoms at first may be mild, serious breathing problems may follow later if the exposure is high enough.

Gases and vapours can produce harmful effects in other ways. Some can cause asphyxiation (deprivation of oxygen to the tissues), and after entering the body through the lungs they can cause damage to other tissues of the body (see below).

13.3.2 Asthma

Asthma is a condition characterized by inflammation of the lining of the airways and intermittent spasm of the underlying smooth muscle. Comparatively more is known about the causes of asthma through work exposure ('occupational' asthma) than about other forms. It is often, but not invariably, the result of allergy to an inhaled dust or vapour in the workplace. Its symptoms include cough, wheeze, chest tightness, and shortness of breath which usually improve on days off work or longer holidays but the association with work may be difficult to establish in some cases. In the UK there are probably more than 2000 new cases every year, and there have been some fatilities.

Important causative agents include diisocyanates (*e.g.* in twin-pack spray paints), hardening/curing agents, *e.g.* anhydrides, rosin (colophony) fumes from soldering flux, aldehydes, *e.g.* formaldehyde or glutaraldehyde, cyanoacrylates (as in 'superglue'), and antibiotics. Causative agents of larger molecular weight include dusts from various cereals (including flour), and urinary protein from mammals (rats, mice), and even allergens from insects and crustaceans.

In the home, exposure to allergens from house dust mites can be a contributing factor in the development of asthma as well as a cause of its symptoms. Other allergens from pollen, moulds, animal dander, *etc.* can cause asthmatic symptoms. Outside the home, in the general environment, increase in asthmatic symptoms has been attributed to exposure to soya bean dust and to oil seed rape. The contribution to the causation of asthma by irritant gases such as sulfur dioxide, nitrogen dioxide, and ozone is still unclear, although it is known that these substances may aggravate symptoms in those who are already asthmatic.

13.3.3 Chronic Bronchitis

The best documented and most important environmental cause of chronic bronchitis is tobacco smoke. Many substances (such as sulfur dioxide) can aggravate the symptoms of bronchitis and cause premature deaths from this condition, as occurred in the smogs that affected many big cities in the early 1950s.

13.4   DISEASES MAINLY OF THE RESPIRATORY UNITS

13.4.1 Pneumoconiosis

Pneumoconiosis is the non-neoplastic (*i.e.* excludes cancer) reaction of the lungs to inhaled mineral or organic dust and results in an alteration of lung structure and deterioration of function. It also excludes diseases mainly of the airways like asthma and bronchitis. Two important pneumoconioses are coal workers' pneumoconiosis and silicosis.

13.4.1.1 Coalworkers' Pneumoconiosis (cwp). This is a pneumoconiosis caused by inhalation of coal dust and is more prevalent with longer exposures to, or with higher concentrations of, dust. The lung is damaged by fibrosis (scarring) and emphysema (destruction of respiratory units, usually leaving functionless empty spaces).

13.4.1.2 Silicosis. Silicosis is a pneumoconiosis caused by inhalation of quartz (or some other crystalline forms of silicon dioxide), which is lethal to macrophages that ingest it and releases their enzymes. In its early stages it is similar to cwp but the nodules in the lung tend to be denser. It is a serious and progressive disease. The term mixed-dust fibrosis describes the pulmonary disorder caused by the inhalation of silica dust simultaneously with another non-fibrogenic dust. Other mineral pneumoconioses may be caused by beryllium, talc, kaolin, and mica.

13.4.2 Asbestosis

Asbestosis is often classified separately from pneumoconiosis for, though asbestos is a dust, it is a special form of fibrous dust. Like silicosis, asbestosis is a serious condition which is incurable and can result in death at an early age. However, as is the case with many harmful substances it does require a certain inhaled dose of asbestos before there is a measurable risk. Thus, living or working in a building in which asbestos has been used in construction but in which the asbestos has been adequately sealed, will not result in a high enough inhaled dose, and cannot cause asbestosis. (*Note*: the term 'asbestosis' is only used for one specific disease caused by asbestos and not for all its health effects.) Other effects linked to asbestos exposure can include thickening and/or calcification of the pleura. (Asbestos can also cause a special cancer of the pleura called mesothelioma as well as bronchial cancer similar to that caused by smoking; *see* below.)

13.4.3 Extrinsic Allergic Alveolitis

Extrinsic allergic alveolitis is an inflammation of the respiratory units of the lung caused by specific inhaled allergens. It can be caused by sensitization to many organic dusts, mainly fungal spores, *e.g.* farmer's lung and malt worker's lung. Besides tending to affect the respiratory units of the lung rather than the conducting airways, it may have 'flu'-like symptoms in addition. In some respects it is similar to humidifier fever which may be caused by sensitization to amoebae or algae. Inhalation of oil mists may cause lipid pneumonia (or other conditions such as asthma) depending on their composition.

13.4.4 Nuisance Dusts

So-called 'nuisance dusts' are relatively inert and, by definiton, cause no serious health effects, although they may be irritant to the upper airways. Examples include chalk, limestone, and titanium dioxide. They may cause radiographic changes without disease. Dusts should be considered as nuisance dusts only when there is good evidence that they are inert and free from significant health effects, not when evidence for an effect is lacking because it has not been carefully sought. Some studies indicate that ultrafine particulate matter such as may be formed by combustion or condensation may be absorbed through the alveoli, be ingested by macrophages and, through cascade mechanisms, result in inflammation that may damage the lungs or other organs. Thus, while titanium dioxide dust is usually regarded as non-toxic, ultrafine titanium dioxide has been shown to cause serious inflammation in animal experiments.

13.5   CANCERS

The single most important known environmental respiratory carcinogen by far, in man, is tobacco smoke. However lung cancer may also be caused by other agents, *e.g.* asbestos, certain compounds of nickel, polycyclic aromatic hydrocarbons (*e.g.* benzpyrene), arsenic trioxide, and chromates. The genesis of cancer is a complex, multistage process and various toxic agents may contribute to its development. Thus, for example, a heavy tobacco smoker might have a risk of developing lung cancer which is about 10 times that of a non-smoker. A worker with heavy asbestos exposure may have about a 5-fold increase in risk of lung cancer compared with a non-exposed worker. If the heavily asbestos exposed worker also smoked heavily then the 'relative' risk of developing lung cancer might be approximately 50-fold.

Exposure to asbestos dusts, probably of all kinds, but especially of blue asbestos (crocidolite), causes mesothelioma which is a cancer of the pleural lining of the lung (besides also causing an increased risk of lung cancer in the bronchus as with smokers). Hundreds of asbestos ex-workers still die of these diseases in the UK every year.

Cancer of the nose or nasal sinuses might be caused by inhalation of certain dusts from hard woods, and from nickel refining.

13.6   SYSTEMIC EFFECTS

13.6.1 Dusts

Some dusts, *e.g.* lead and its salts, can be absorbed into the body after inhalation or skin contact. They can then have harmful effects on other organs, *e.g.* the nerves or the blood-forming organs.

13.6.2 Systemically Toxic Gases and Vapours

Methylene chloride, various chloroethanes, and chloroethylenes are examples causing systemic toxicity. The effect of methylene chloride, although similar to that of vapours given off by other organic solvents such as trichloroethylene, presents the added hazard of being metabolized by the body to carbon monoxide. Initially these might cause a feeling of well-being similar to that produced by alcohol. At higher concentrations they cause unconsciousness. Repeated exposure can lead to permanent brain damage.

13.6.3 Simple Asphyxiant Gases

Life depends on an adequate supply of oxygen reaching the tissues of the body. Oxygen present in the air breathed into the lungs passes into the blood and is carried to the tissues. Simple asphyxiants may interfere with this process by displacing oxygen from the air that is breathed in (examples are methane and nitrogen). This happens usually in enclosed, poorly ventilated spaces particularly underground where methane can be produced by naturally occurring processes or where natural oxygen has been depleted. Many fatalities have occurred as a result of oxygen depletion being overlooked. Carbon dioxide also causes rapid breathing, headache, sweating, and if exposure is high enough, loss of consciousness, and death.

13.6.4 Chemical Asphyxiant Gases

These cause asphyxia by interfering with oxygen transport (examples are carbon monoxide, hydrogen cyanide, and hydrogen sulfide). Carbon monoxide is present when there is incomplete combustion of carbon-containing fuels. It is odourless. Hydrogen sulfide occurs in mines, sewers, and wherever there are chemical processes involving sulfur compounds such as in slurry pits. It may be detected by its smell at low concentrations, but this cannot be relied upon as a warning since it paralyses the sense of smell and can effectively become odourless. Hydrogen cyanide in the form of its salts (cyanides), is used in electroplating and other industries, and organic cyanides (nitriles) of varying degrees of toxicity are also used in industry. Absorption can also occur through the skin. At low concentrations these gases cause the rapid onset of headache, dizziness, vomiting, and confusion. At higher concentrations they are very rapidly lethal.

13.7 BIBLIOGRAPHY

C.D. Klaassen, ed. 'Casarett and Doull's Toxicology', McGraw-Hill, New York, 5th edn, 1996.

W.K.C. Morgan and A. Seaton. 'Occupational Lung Diseases'. Saunders, Philadelphia, 3rd edn., 1995.

W.R. Parkes. 'Occupational Lung Disorders', Butterworth, London, 2nd edn, 1994.

P.A.B. Raffle, P.H. Adams, P.J. Baxter, and W.R. Lee, eds. 'Hunter's Disease of Occupations'. Edward Arnold, London, 8th edn, 1994.

CHAPTER 14

Hepatotoxicity

ALISON L. JONES

14.1   INTRODUCTION

Many potentially toxic substances enter the body *via* the gastrointestinal tract (gut). As the blood supply from the gastrointestinal tract (through the portal vein) drains into the liver, the liver comes into contact with the potentially toxic substances, and this exposure will often be at a higher concentration than in other tissues.

The liver is essential for the metabolic disposal of virtually all xenobiotics (foreign substances). This process is mostly achieved without injury to the liver itself or to other organs.

A few compounds such as carbon tetrachloride are toxic themselves or produce metabolites that cause liver injury in a dose-dependent fashion.

However, most agents cause liver injury only under special circumstances when toxic substances accumulate. Factors contributing to the build-up of such toxic substances include genetic enzyme variants (enzymes with altered function due to gene defects), which allow greater formation of the harmful metabolite, and induction (greater production) of an enzyme which produces more than the usual quantity of toxic substance. There may also be accumulation of toxic substances by interference with regular non-toxic metabolic pathways by substrate competition for enzymes (*e.g.* ethyl alcohol and trichloroethylene) or depletion of substrates used to metabolize the toxins.

14.2.   THE ANATOMY OF THE LIVER

The liver is served by two blood supplies, the portal vein, which delivers 75% of the blood supply and the hepatic artery, which delivers the rest. The portal vein drains the gastrointestinal tract, spleen, and pancreas and supplies nutrients and some oxygen. The hepatic artery supplies fully oxygenated blood to the liver.

The blood that enters the liver by the hepatic artery and portal vein flows through sinusoids (Figure 14.1). Sinusoids are specialized capillaries with large pores which allow large molecules to pass through into 'interstitial space' and into close contact with hepatocytes (liver cells); therefore foreign compounds are taken up very readily by hepatocytes. The liver is composed of hepatocytes arranged as plates: each plate is bounded by a sinusoid. The membranes of adjacent hepatocytes form the bile caniculi (tubes) into which bile is secreted. The bile caniculi form a network and feed into ductules which become bile ducts (Figure 14.1). Bile serves to excrete compounds from the body and aids in the digestion of food in the gastrointestinal tract.

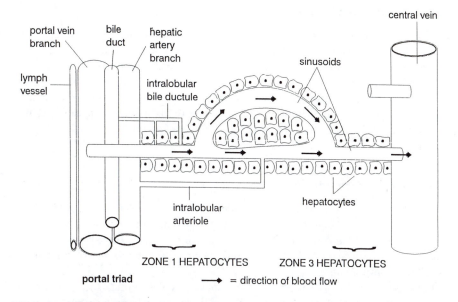

Figure 14.1 *The anatomy of the liver*
(This figure has been adapted with kind permission of the publisher from Figure 7.6 in J.A. Timbrell, 'Principles of Biochemical Toxicology'. Taylor & Francis, London, 2nd edn., 1992)

Hepatocytes near the portal tract branches (zone 1) receive blood that is rich in oxygen and nutrients but those near the hepatic vein branches (zone 3) receive blood that has lost much of its nutrients and oxygen. Therefore zone 3 is particularly sensitive to damage from toxic compounds. Zone 3 cells have a higher level of some metabolic enzymes and higher lipid synthesis than zone 1 which may also explain why zone 3 tends to be the most damaged and why lipid accumulation is a common response to this damage (*see* the carbon tetrachloride example below).

Allyl alcohol (2-propen-1-ol) however causes zone 1 necrosis partly because this is the first area exposed to the compound in the blood and partly as a result of the presence of alcohol dehydrogenase in zone 1 which produces reactive metabolites.

14.3 MECHANISMS OF CELLULAR INJURY

Toxic substances can damage cells in target organs in a variety of ways. The eventual pattern of response may be reversible injury or an irreversible change leading to the death (necrosis) of the cell or perhaps to carcinogenesis (cancer).

The toxic substance (or its metabolites) may cause cell damage by a variety of mechanisms (one or several of which may occur), some of which are considered below.

14.3.1 Covalent Binding

As well as free radicals, other reactive intermediates may be produced by metabolism. These can interact with proteins and other macromolecules binding covalently to them. Studies have demonstrated a correlation between the amount of binding and tissue damage, though this may merely reflect production of other damaging species. Binding to critical sites on proteins could, however, alter their function by, for example, inhibiting an enzyme or damaging a membrane, but binding could be to non-critical sites, and therefore be of no toxicological importance.

14.3.2 Lipid Peroxidation

Free radicals arise by cleavage of a covalent bond in a molecule, the addition of an electron or by the taking of a hydrogen atom from another radical. They have an unpaired electron centred on a carbon, nitrogen, sulfur, or oxygen atom and hence are extremely reactive, electrophilic species which can react with a variety of cellular components.

Lipid peroxidation is caused by the attack of a free radical on unsaturated lipids (particularly polyunsaturated fatty acids found in membranes), the reaction being terminated by the production of lipid alcohols, aldehydes, or malondialdehyde. Therefore there is a cascade of peroxidative reactions which leads to the destruction of lipid unless stopped by a protective mechanism or a chemical reaction such as disproportionation which gives rise to a non-radical product.

The structural integrity of membrane lipids will be adversely affected leading to alterations in fluidity or permeability of membranes, destabilisation of lysosomes (intracellular organelles), and altered function of the endoplasmic reticulum and mitochondria. Such mechanisms are thought to be involved in liver damage caused by carbon tetrachloride and white phosphorus.

14.3.3 Thiol Group Changes

Glutathione is responsible for cellular protection. If depleted, a cell is made more vulnerable to toxic substances. Reactive intermediates of toxic substances can react with glutathione either by a direct chemical reaction or by a glutathione transferase-mediated reaction. If excessive, these reactions can deplete cellular glutathione and leave essential proteins vulnerable to attack by oxidation, cross-linking, formation of disulfides, or covalent adducts.

14.3.4 Enzyme Inhibition

Sometimes inhibition of an enzyme may lead to cell death, for example cyanide inhibits cytochrome aa_3 leading to blockage of cellular respiration. This results in depletion of adenosine triphosphate (ATP) (ATP is produced by mitochondria and is the main energy source within the cell) and other vital endogenous molecules.

14.3.5 Ischaemia

Reduction of oxygen or nutrients supplied to cells results in cell damage and eventual cell death if prolonged. Ischaemia may be a secondary event due to swelling of cells with reduction of blood flow. For example—phalloidin causes centrilobular necrosis (see under patterns of response to injury in the liver).

As a result of such mechanisms of damage as described above and the inability of the cell to compensate for changes in membrane structure and permeability, changes in subcellular skeleton and organelles may occur. This may result in ATP (energy) depletion, changes in Ca^{2+} concentration, damage to intracellular organelles, and DNA (deoxyribonucleic acid) damage, and stimulation of apoptosis (programmed cell death) may occur.

14.3.6 Depletion of ATP

This may be caused by many toxic substances, usually by the uncoupling of mitochondrial oxidative phosphorylation or by DNA damage which causes activation of poly(ADP–ribose) polymerase (where ADP = adenosine 5'-diphosphate). Depletion of ATP in the cell means that active transport in and out of the cell is altered or stopped and changes in electrolytes, particularly Ca^{2+}, lead to changes in biosynthesis within the cell such as protein synthesis, production of glucose and lipid synthesis.

A very important mechanism of cellular damage is alteration of Ca^{2+} concentration. Changes in the intracellular distribution of this ion have been implicated in the cytotoxicity of many toxic substances including carbon tetrachloride. Interference with Ca^{2+} homeostasis may occur as a result of inhibition of Ca^{2+} adenosine triphosphatases (ATPases which are responsible for the transport of Ca^{2+} across cell membranes), direct damage to the plasma cell membrane allowing leakage of Ca^{2+}, or depletion of intracellular ATP.

14.3.7 Damage to Intracellular Organelles

Damage to intracellular organelles can result from the above mechanisms of injury, for example carbon tetrachloride damages both smooth and rough endoplasmic reticulum leading to disruption of protein synthesis of the whole cell. Mitochondrial damage may occur, for example, after exposure to hydrazine, leading to functional changes and rupture of mitochondria. The mitochondria are crucial to the cell, and inhibition of their electron transport chain leads to rapid cell death.

14.3.8 DNA Damage

DNA damage may result from compounds such as alkylating agents, *e.g.* dimethyl sulfate can cause single-strand breaks in DNA resulting in the activation of poly(ADP–ribose) polymerase which catalyses post-translational protein modification and is involved in polymerization reactions and DNA repair. Severe DNA damage may result from its activation and be sufficient to lead to cell death or carcinogenesis.

14.3.9 Apoptosis

Apoptosis is programmed cell death. Some foreign compounds may stimulate such cell death by, for example, the influx of calcium into a cell.

14.4   PATTERNS OF RESPONSE TO INJURY IN THE LIVER

The resulting pattern of liver damage from the above mechanisms reflects the fact that the liver's response to injury is limited and includes fatty liver, necrosis (cell death), cholestasis, cirrhosis, and carcinogenesis.

14.4.1 Fatty Liver

This is the accumulation of triglycerides (fats) in the liver cells. Fatty liver is a common response to toxicity, often occurring as a result of interference with protein synthesis in hepatocytes, such as after exposure to hydrazine (NH_2-NH_2) for example. Normally it is a reversible process which does not lead to cell death although it can occur in combination with liver cell death (necrosis) as is the case with carbon tetrachloride exposure.

Repeated exposure to compounds that cause fatty liver, such as ethyl alcohol, may lead to cirrhosis.

14.4.2 Necrosis (Cell Death)

This may occur by direct cell injury, with disruption of intracellular function, or by indirect injury by immune-mediated membrane damage. As mentioned previously, allyl alcohol causes periportal (zone 1) necrosis partly because alcohol dehydrogenase is present in zone 1 and partly because this is the first area exposed to the compound in the blood.

Conversely carbon tetrachloride and bromobenzene cause zone 3 (centrilobular) necrosis as a result of metabolic activation in that region. The hepatotoxicity of carbon tetrachloride has been extensively studied. It is a simple molecule which, upon exposure, is distributed widely throughout the body, but despite this, its major toxicity is to the liver because it is dependent on metabolic activation by the P_{450} cytochrome system (a haem-like molecule with maximum absorbance at 450 nm: previously called mixed-function oxidase). Carbon tetrachloride is metabolically activated by dechlorination to yield a free radical as shown in Figure 14.2. The carbon tetrachloride first binds to cytochrome

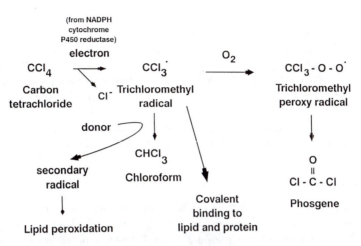

Figure 14.2 *The metabolic activation of carbon tetrachloride*
(This figure has been adapted with kind permission of the publisher from Figure 7.3 in J.A. Timbrell, 'Principles of Biochemical Toxicology'. Taylor & Francis, London, 2nd edn., 1992)

P_{450} and then receives an electron from NADPH (reduced form of nicotinamide–adenine dinucleotide phosphate) cytochrome P_{450} reductase. The enzyme–substrate complex then loses a chloride ion and the free radical intermediate is generated. This may then react with oxygen or take a hydrogen atom from a suitable donor to yield a secondary radical, or react covalently with lipids or protein.

Chronic exposure causes liver cirrhosis or tumours, particularly in the centrilobular area which contains the greatest concentrations of P_{450} cytochrome. The production of reactive metabolites appears to start a cascade of damage, such as lipid peroxidation. Low-dose exposure may cause only fatty liver.

Mid-zonal (zone 2) necrosis is less common than the other two types of necrosis, but occurs with beryllium toxicity.

The explosive trinitrotoluene (TNT) can cause massive liver necrosis involving all zones. Ischaemia (impaired blood supply to the liver) may also contribute to necrosis, for example phalloidin, a toxic substance present in poisonous mushrooms, may cause swelling of the cells lining the sinusoids (Figure 14.1) and therefore reduce the oxygen and nutrients supplied to hepatocytes.

14.4.3 Cholestasis (Impaired Bile Flow)

Bile duct injury may result from exposure to a number of compounds, particularly those which are concentrated in bile. The result of the damage will be cholestasis due to debris from necrotic cells blocking the ductules. Because of the close interrelation between bile ducts and hepatocytes this may also be accompanied by damage to hepatocytes (*e.g.* α-naphthyl isothiocyanate) by the build-up of bile which damages cell membranes in excess.

14.4.4 Cirrhosis

Cirrhosis is characterized by regenerative nodules (clumps of new cells) within fibrotic tissue (amorphous tissue) forming an irregular lobulated (defined) pattern. Any repetitive injury resulting in cell death (necrosis) followed by repair mechanisms may lead to cirrhosis. This is because the liver has only a limited capacity to regenerate.

In addition, compounds that do not appear to cause acute necrosis such as ethionine and alcohol may cause cirrhosis after chronic exposure.

14.4.5 Carcinogenesis

Liver tumours may be benign (grow *in situ*) or malignant (able to metastasize to other tissues). They may arise from any cell type within the liver, *e.g.* adenoma (aflatoxin B₁), hepatocellular carcinoma (dimethylnitrosamine), haemangiosarcoma (vinyl chloride).

Among the various mycotoxins (toxins produced by moulds on nuts, oil seeds, and grains) the aflatoxins have been the subject of intensive research because they are potent carcinogens. Aflatoxin B_1 is a very reactive compound. Its carcinogenicity is associated with its biotransformation to a highly reactive, electrophilic oxide which forms covalent adducts with DNA, ribonucleic acid (RNA), and protein (Figure 14.3). Damage to DNA is thought to induce tumour growth. There may be species differences due to differences in biotransformation and susceptibility to the initial biochemical lesion.

Dimethylnitrosamine is evenly distributed throughout the body but exposure to single doses causes centrilobular hepatic necrosis indicating that metabolism is an important factor in its toxicity. One metabolite is a highly reactive alkylating agent that will methylate nucleic acids and proteins. The degree of methylation of DNA *in vivo* correlates with the risk of tumour induction in those tissues.

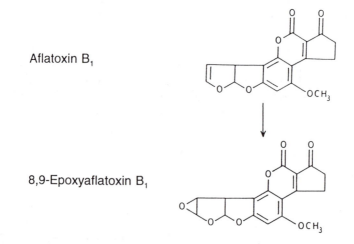

Figure 14.3 *Biotransformation of aflatoxin B₁*

Vinyl chloride (or vinyl chloride monomer) is the starting point in the manufacture of poly(vinyl chloride). Chronic exposure leads to a 'vinyl chloride disease' which includes skin changes, changes to the bones of the hands, and liver damage.

Haemangiosarcoma is a tumour of sinusoidal cells (not hepatocytes) that may also result from chronic exposure to vinyl chloride. This again appears to occur because the epoxide intermediate and fluoroacetaldehyde bind to DNA and proteins respectively within the cell. Haemangiosarcoma has also been associated with arsenite (AsO_2^-) exposure although the mechanism of this is still unclear: it should be stressed that the experience in humans with carcinogenicity has not been replicated in animal models.

14.4.6 Veno-occlusive Disease

Rarely, a toxin may damage sinusoids and endothelial cells directly, for example monocrotaline, a plant alkaloid, which is metabolized to a reactive molecule causing damage and blockage of the venous return to the liver and secondary ischaemic death of hepatocytes.

14.4.7 Proliferation of Peroxisomes (Micro Bodies)

Peroxisomes are organelles found in many cell types but particularly in liver cells, and they take part in the metabolism of lipids (fats) predominantly. Exposure to certain substances leads to an increased number of peroxisomes and this may lead to an unwanted increase in the activities of various degradative enzymes.

14.5   DETECTION OF LIVER DAMAGE

Measurement of the plasma activities of the enzymes aspartate transaminase, alanine transaminase, alkaline phosphatase, and γ-glutamyl transpeptidase (GGT) is a common way of detecting liver damage in live animals or humans. They are intracellular enzymes that are released when liver cell death occurs. Measurement of the plasma activity of alkaline phosphatase or GGT may reflect damage to biliary cells as they contain large quantities of these enzymes. Liver function may be estimated by measuring the hepatic clearance (uptake from blood) of a dye such as sulfobromophthalein, although this is outmoded now as a routine biochemical investigation of liver disease.

In pathology, damage to the liver can be detected by light or electron microscopy of liver sections. In animals or humans liver biopsy may be undertaken to look at the pattern of liver injury. It is of limited value unless the potentially toxic substances are known, as the pattern of response to injury is limited within the liver.

14.6   CLINICAL PROBLEMS RESULTING FROM LIVER DAMAGE

Liver damage may be silent or the following may occur:

- Jaundice. Excess quantities of bilirubin (a pigment produced by the liver from the degradation of haem in the red blood cells) is an indicator of cell death or dysfunction. The individual has yellow skin and eyes which feel itchy.
- Portal hypertension. This is due to cirrhosis, and is characterized by an enlarged spleen and diversion of blood flow from the liver to other areas. In advanced cases individuals may vomit blood or have marked fluid retention causing swelling of the abdomen (ascites).
- Clotting difficulties. The liver is an important producer of proteins for the body. One of its most important functions is the production of clotting proteins making up the clotting cascade. Therefore clotting difficulties may result from liver cell necrosis or cirrhosis and are seen as bleeding or bruising from the skin, gums, or gastrointestinal tract.
- Other features. In humans, anorexia, nausea, or vomiting may be noted in the presence of liver damage.

In general clinical features in humans do not correlate particularly well with type of liver damage and thus biopsies may need to be undertaken to establish the pattern of liver injury. However, in general an individual with a fatty liver may not have symptoms and may be discovered only by increased plasma activities of aspartate transaminase or alanine transaminase (ALT). Necrosis may be asymptomatic if involving only a few liver cells, although more extensive damage may lead to marked jaundice, clotting difficulties, and coma. Cirrhosis may be manifested by the features of portal hypertension. Tumours may present with jaundice and weight loss.

14.7   BIBLIOGRAPHY

N. Kaplowitz, 'Drug metabolism and hepatotoxicity', in 'Liver and Biliary Diseases', ed. N. Kaplowitz, Williams and Wilkins, Baltimore, 1991, pp. 82–97.

C.D. Klaassen, ed., 'Casarett and Doull's Toxicology', McGraw-Hill, New York, 5th edn, 1996.

M. Ruprah, T.K.G. Mant, and R.J. Flanagan, Acute carbon tetrachloride poisoning in 17 patients, *Lancet*,1985, **1**, 1027–1029.

S. Sherlock and J. Dooley, 'Diseases of the Liver and Biliary System'. Blackwell, Oxford 9th edn., 1993. (Recommended chapters: ch. 6: Hepatocellular failure, pp. 72–85; ch. 7: Hepatic encephalopathy, pp. 86–101, ch. 8: Fulminant hepatic failure, pp. 102–113; ch 18: Drugs in the liver, pp. 322–356).

J.A. Timbrell, 'Principles of Biochemical Toxicology'. Taylor & Francis, London, 2nd edn, 1992.

H.J. Zimmerman, and W.C. Maddrey, 'Toxic and drug induced hepatitis', in 'Diseases of the Liver'. Lippincot, New York, 1993, 707–783.

CHAPTER 15

Nephrotoxicity

JOHN H. DUFFUS and ROBERT F.M. HERBER

15.1   INTRODUCTION

Nephrotoxicity was defined in 1991 by the World Health Organization (WHO) as renal disease or dysfunction that arises as a direct or indirect result of exposure to medicines, and industrial or environmental chemicals.

Kidneys exist in pairs and have the dual function of control of fluid and electrolyte balance and control of blood pressure. Thus, they have a major role in total body homeostasis and in the homeostasis of the cardiovascular system.

The kidney is divided into two main areas, the outer cortex and the inner medulla (Figure 15.1). The cortex forms the major part of the kidney and receives most of the blood supply and thus both the nutrients and toxicants that it may contain.

15.2   PARTS OF THE NEPHRON AND THEIR FUNCTIONS

The nephron (Figure 15.2) is the functional unit of the kidney, responsible for filtration of the blood to form urine. The vascular element of the nephron delivers waste products for excretion, returns reabsorbed or synthesized materials to the systemic circulation, and delivers oxygen and metabolic substrates to the nephron. Selective filtration is based on molecular size, shape, and net charge. Ultrafiltration takes place in the glomerulus and, under normal conditions, the glomerular filtrate contains all the constituents of blood except for blood cells and plasma proteins. Pores of about 0.1 μm in diameter allow the passage of filtrate under pressure caused by the efferent blood vessels of the glomerulus being narrower than the afferent vessels.

The tubule of the nephron is responsible for water, glucose, and electrolyte reabsorption or loss, acid–base balance (by production of hydrogen carbonate), and synthesis of ammonia. The filtrate, which is modified in its flow along the tubule, eventually emerges from the kidney as urine. The loop of Henle, by active transport of sodium chloride, produces an osmotic gradient which causes water to be removed from the ultrafiltrate as it passes so that 99% of the original fluid is removed to produce a concentrated urine.

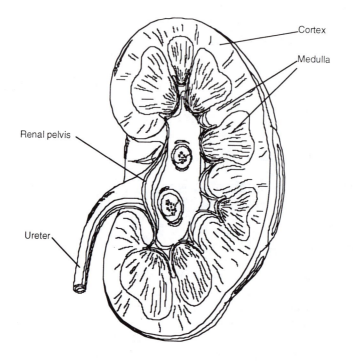

Figure 15.1 *Cross-section of the kidney*

15.3   ENDOCRINE FUNCTIONS OF THE KIDNEY

The kidney is also the site of synthesis of hormones and certain vasoactive prostaglandins and kinins that influence the rest of the body.

Renin, which stimulates the formation of angiotensin and aldosterone, which in turn enhance sodium reabsorption and potassium excretion in the distal tubule, is formed in the kidney, as is angiotensin. Angiotensin causes peripheral vasoconstriction and stimulates aldosterone secretion.

The cortical tubules are responsible for activation of vitamin D precursors. 25-Hydroxyvitamin D undergoes hydroxylation specifically in the kidney to 1,25-dihydroxyvitamin D, which plays a key role in promoting bone resorption and calcium absorption from the gut. Erythropoietin, which stimulates erythrocyte formation in the bone marrow, is also synthesized in the kidney

The distal tubule of the kidney secretes kallikrein, an enzyme that is essential for the production of bradykinin, a very powerful vasodilator which increases the permeability of postcapillary venules. The medullary interstitium and mesangial cells secrete prostaglandins. Prostaglandins cause vasodilation and stimulate diuresis.

The proximal tubules are the site of degradation of insulin, glucagon, and parathyroid hormone.

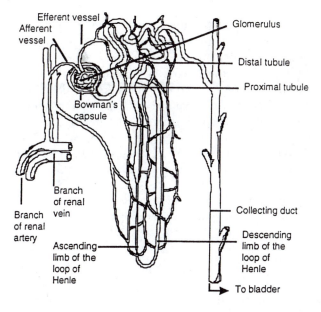

Figure 15.2 *Diagram of the kidney nephron*

15.4 BASIS OF KIDNEY VULNERABILITY TO TOXIC INJURY

The two kidneys comprise about 1% of the body mass but receive about 25% of the cardiac output. About one-third of the plasma water that reaches the kidney is filtered. Maintenance of renal function requires delivery of large quantities of oxygen and metabolic substrates. Thus, the kidney is highly vulnerable to substances that cause anoxia.

Chemicals that damage the kidney may be referred to as nephrotoxicants or, if they are of biological origin, nephrotoxins. Nephrotoxicity is defined as the pathological state in which the normal homeostatic function of the kidney is disrupted. It is usually associated with necrosis of one or more segments of the proximal tubule. The disruption of renal function may present itself either with the excretion of large volumes of very dilute urine (high output renal failure) or with the excretion of minimal amounts of urine (anuria).

An important determinant in kidney toxicity is the balance between the generation of reactive metabolites and the rate of their removal. As in the liver, reactive metabolites are often strong electrophiles or free radicals which cause necrosis by binding to macromolecules or initiating lipid peroxidation. Among the toxicants which act in this way are paracetamol (acetaminophen), bromobenzene, chloroform, and carbon tetrachloride.

The kidney is a highly active organ metabolically and it has a very high rate of oxygen consumption. Several renal transport mechanisms in the tubule can concentrate xenobiotics in the tubular epithelial cells. Reabsorption of cationic compounds from the tubular fluid, *e.g.* gentamicin, may lead to nephrotoxic concentrations of this compound in the epithelial cells of the proximal tubule. Cadmium metallothioneins are absorbed actively into the proximal cells. After the degradation of the metallothionein by lysosomes, cadmium is released into the cytosol (cell fluid) and thus exerts its toxicity. The kidney filters and reabsorbs large quantities of water and concentrates potentially toxic substances in the renal tubule lumen. It is exposed to high concentrations of foreign chemicals by glomerular filtration. Further, this is greatly influenced by active tubular secretion, tubular reabsorption, and by the counter-current mechanism in the loop of Henle. In addition, there is, because of the rich vascularization, a large endothelial surface area for the deposition of antigen–antibody complexes.

There are some protective mechanisms against toxic substances in the kidney but qualitatively they do not differ much from those in other organs. They include:

- Cytochrome P450-dependent oxidations
- Glutathione reduction
- Epoxide hydrase hydrolytic reactions
- Glucuronide formation
- Protein complexes (metallothionein)
- Lysosomes (hydrolytic enzymes)

Biotransformation of the parent compound to a toxic metabolite may be a factor in nephrotoxicity. Although the kidney does not have levels of xenobiotic-metabolizing enzymes (*e.g.* the cytochrome P450-dependent monooxygenase system) as high as those in the liver, some of the same enzymic reactions have been shown to occur in the kidney. The concentration of cytochrome P450 is highest in the cells of the pars recta of the proximal tubule, an area which is particularly susceptible to toxic damage. It is probable that covalent binding to tissue macromolecules occurs in close proximity to the site of activation. Thus, chemicals that cause damage through reactive intermediates are probably activated in the kidney rather than being activated in the liver and then being transported to the kidney.

15.5   NEPHROTOXICANTS

Occupational or environmental exposure to chemicals may cause renal effects. Most nephrotoxic chemicals in the working and general environment are either metals or volatile hydrocarbons. A primary nephrotoxicant is one that induces nephrotoxicity without bioactivation, for example, a mercury(II) salt. A secondary nephrotoxicant is one that induces nephrotoxicity following biotransformation to an ultimate toxic species, for example bromobenzene, trichloroethene.

Important groups of substances that are frequently identified as nephrotoxicants are metals (*e.g.* cadmium, mercury, lead), mycotoxins, drugs (antibiotics, analgesics, antineoplastics), organic solvents, and pesticides.

15.5.1 Metals

Metals cause nephrotoxicity in various ways. Some form complexes or chelate with organic ligands, *e.g.* mercury binds to sulfhydryl groups. Others substitute for endogenous substrates making normal metabolic processes impossible, *e.g.* lead may substitute for calcium, and arsenate for phosphate. Still others cause interference with transport systems or reabsorption processes, *e.g.* chromate competes with sulfate for transport into the cell, and cadmium binds to metallothionein, preventing reabsorption of essential metal ions with an oxidation state of II.

Metals enter proximal tubular cells by endocytosis following the binding of the metal or a metalloprotein complex to the brush-border membrane. Entry is followed by intracellular release of the metal from the membrane or protein–metal complex by lysosomal degradation. Low doses of certain metals cause leaking of glucose and amino acids (renal tubular acidosis) associated with increased diuresis. Then renal tubular necrosis occurs and this may lead to renal shutdown, raized blood urea, and finally death. Necrosis is thought to be due to a combination of ischaemia, caused by vasoconstriction, and direct cytotoxic action.

The early effects of occupational exposure to inorganic or elemental mercury are the excretion in urine of proteins (both low and high molecular mass), enhanced activities of the enzymes N-acetyl-β-D-galactosidase (NAG) and γ-glutamyl transferase, and tubular antigens. Mercury(II) chloride causes both vasoconstriction and cytotoxicity in the proximal tubule. Inorganic mercury ions may also produce an immunologically mediated glomerulonephritis. In both cases, increased concentrations of albumin may be detected in the urine.

The relative nephrotoxicity of different organomercurials broadly reflects their rate of decomposition to release mercury ions. Thus, as with inorganic mercury compounds, symptoms may include polyuria and excretion of albumin, but they may be absent following exposure to very stable mercurials such as methylmercury. Results of studies of exposure of humans to methylmercury *per se* are not clear as far as effects on the kidney are concerned. Phenylmercury may cause reversible damage to the tubules of children exposed to this fungicide.

Dysfunction of the proximal renal tubules is characteristic of chronic but not acute cadmium exposure in both humans and experimental animals. Acute doses of cadmium accumulate mainly in the liver but chronic exposure leads to accumulation in the kidney. Cadmium–protein complexes accumulated by hepatic cells induce the synthesis of metallothionein, a low molecular mass cysteine-rich protein which binds cadmium avidly. This cadmium–metallothionein complex is thought to be very gradually released from the liver and taken up by endocytosis into renal tubule cells, where free cadmium ions can be generated owing to lysosomal degradation of the complex.

The critical cadmium concentration when kidney damage occurs is about 100 $\mu g\ g^{-1}$ of renal cortex. Cadmium in the kidney also induces the synthesis of metallothionein in the proximal tubule, and dysfunction probably reflects saturation of the available binding sites. An early effect on the kidney of chronic exposure is the excretion in urine of the low molecular mass proteins such as β_2-microglobulin, α_1-microglobulin and retinol binding protein, which are all indicators of effects on the tubule. Also there is increased activity in urine of some tubular enzymes, for example NAG. Claims have been made also that there is an effect on the glomerulus, but this remains controversial. Prolonged exposure at higher concentrations may lead to a diminished glomerular filtration ratio (see Figure 15.3) and a decreased kidney function.

Acute necrosis of the proximal convoluted tubule has been reported following exposure to chromium(VI), the effect being similar to that of inorganic mercury ions. Results of studies of exposure of humans to low chronic doses of chromium show inconsistent kidney effects. Sometimes increased urinary excretion of β_2-microglobulin and total protein are seen and sometimes the activity of tubular β-glucuronidase is raised.

Probably the most abundant nephrotoxic metal is lead. Nephrotoxicity, however, is observed only after relatively high occupational exposure giving a blood concentration of 600 μg l^{-1} or more. Lead binds to high-affinity lead-binding proteins which are present in the kidney tubule cells in high concentrations. These proteins carry lead to the nucleus where *de novo* synthesis of a unique acidic protein results in metal precipitation to form lead intranuclear inclusion bodies.

Mitochondria are extremely sensitive to lead and, following chronic *in vivo* administration, mitochondrial swelling occurs. The mitochondria show impaired oxidative phosphorylation and thus the tubule cells are less able to reabsorb or secrete solutes and metabolites.

An early effect on the kidney of exposure to lead is the higher activity in urine of NAG and the enhanced concentration of the tubular proteins, retinol binding protein and α_1-microglobulin. In a study of lead smelter workers performed in the USA, an excess mortality rate was found due to chronic renal failure.

Other inorganic compounds can cause nephrotoxicity and result in increased urinary protein output and/or tubular enzyme activity. Some examples in humans are uranium (β_2-microglobulin and enzymes), nickel (β_2-microglobulin), arsenic(V) (retinol binding protein), and SiO_2 (albumin and α_1-microglobulin).

15.5.2 Mycotoxins

Mycotoxins may have profound effects on the fundamental biochemistry and physiology of the kidney. For example, Ochratoxin A, a secondary metabolite of Aspergillus and Penicillium species, may be present in contaminated grain products and in meat from animals eating contaminated feeds. In pigs, it produces porcine nephropathy, a disease characterized by degeneration and atrophy of proximal tubules, interstitial fibrosis, and hyalinization of glomeruli. It also produces nephrotoxicity and renal tubule tumours in rats and mice and may cause the disease known as Balkan Endemic Nephropathy in humans.

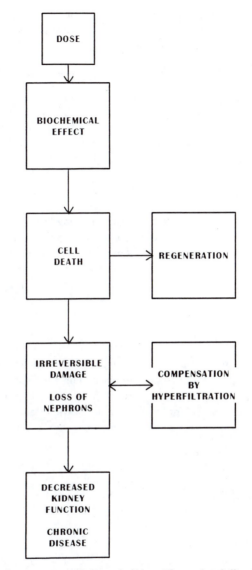

Figure 15.3 *Effect of exposure to xenobiotics on the kidney. The severity of effects increases from top to bottom*

15.5.3 Drugs

Analgesics such as phenacetin, aspirin, indomethacin, and mefenamic acid produce renal papillary necrosis in rats and possibly also in humans. This disease is characterized by necrosis of the endothelial cells of papillary vessels, the thin limbs of the loops of Henle and the collecting ducts, and the renomedullary interstitial cells which synthesize prostaglandins. At least two possible mechanisms have been suggested. One involves inhibition of prostaglandin synthesis leading to vasoconstriction and consequent ischaemia of the renal papilla. The other involves concentration of the compound or its conjugates in the medulla with bioactivation to a reactive radical intermediate which subsequently causes the damage.

15.5.4 Organic Solvents

The nephrotoxicity of some halogenated hydrocarbons such as chloroform depends on the generation of phosgene by cytochrome P450 action. Phosgene is trapped by two molecules of glutathione to form diglutathione dithiocarbamate which undergoes further metabolism to 2-oxothiazolidine 4-carboxylic acid which is excreted. Other haloalkenes may undergo metabolic activation by other routes.

Prolonged exposure to volatile hydrocarbons may lead to effects on the kidney (glomerulonephritis). Early renal effects of exposure to the following compounds have been described: tetrachloroethene, styrene, toluene, and mixtures of solvents. The effects reported are small and sometimes not related to the dose. Other variables such as shift work or heavy physical work may be responsible for the effects.

15.5.5 Pesticides

The herbicides diquat and paraquat are nephrotoxic. Acute kidney insufficiencies have been reported, but there are no studies on chronic exposure in humans. The nematicide 1,3-dichloropropene is suspected of being nephrotoxic in humans, and possibly the *trans* compound is responsible for the nephrotoxicity only.

15.5.6 Renal Carcinogens

Renal cell cancer accounts for about 75% of all kidney cancer in adults. A number of risk factors for renal cell cancer have been identified including the following:

- Aflatoxin B
- Analgesics
- Asbestos
- Oven operation
- Diethylnitrosamine
- Dimethylnitrosamine
- Diuretics
- Printing press operation
- Streptozotocin
- Tobacco products

All of the factors listed have been associated with DNA (deoxyribonucleic acid) damage and mutagenicity.

Hexachlorobutadiene (HCBD) is a by-product of the manufacture of the chlorinated solvents, trichloroethene and perchloroethene. In the rat, HCBD produces tubule necrosis and renal tubule tumours. It conjugates with glutathione in liver forming pentachlorobutadiene glutathione (PCBD-GSH). PCBD-GSH is metabolized to pentachlorobutadiene *N*-acetylcysteine (PCBD-NAcCys) in renal tubule cells. PCBD-NAcCys is deacetylated and then converted to pyruvate, ammonia, and an electrophilic mercaptan species which reacts with thiols, proteins, and nucleic acids.

15.6 DETERMINATION OF THE EFFECT OF CHEMICALS ON THE KIDNEY

Standard clinical tests of kidney function include the following:

 – Urine volume and osmolality
 – Urinary pH
 – Excretion of the electrolytes sodium and potassium
 – Presence of glucose or excess total protein in urine
 – Changes in urinary sediment
 – Plasma electrolytes, urea, and creatinine

More advanced clinical methods are the measurement of:

 – The glomerular filtration ratio
 – The reserve capacity of the kidney

High-resolution proton NMR spectroscopy can provide a screen for detecting abnormal patterns of metabolites in urine. Proximal tubule toxicants produce marked glycosuria, amino aciduria, and lactic aciduria. Papillary toxins cause early increases in trimethylamine *N*-oxide and dimethylamine. Renal clearance of creatinine or inulin may be measured in order to determine glomerular filtration. The measurement of renal clearance and excretion of *p*-aminohippuric acid may be used also to determine renal plasma flow, and from this renal blood flow can be estimated.

Toxic effects of chemicals may be evaluated *in vitro* by adding the chemical or its metabolite directly to renal cortical slices or isolated proximal tubular cells or tubules.

Histopathological examination of kidney tissue following *in vivo* or *in vitro* exposure can identify structural alterations and the areas affected. Light microscopy identifies selective damage to the nephron caused by chromium (pars convoluta), hexachloro-1,3-butadiene (pars recta), and propeneimine (papilla).

All these methods do have the disadvantage that they determine or measure the damage already present (see Figure 15.3). This means that the damage can be irreversible at the moment of measurement. A better method is to determine early kidney effects, *e.g.* NAG, α_1-microglobulin, retinal binding protein and albumin in urine. Measurement of these parameters gives an early indication of effects and thus preventive measures can be taken (*e.g.* occupational or environmental hygienic measures).

15.7   BIBLIOGRAPHY

B. Ballantyne, T. Marrs, and P. Turner, eds., 'General and Applied Toxicology', (2 volumes). Macmillan, London, 1993.

C.D. Klaassen, ed. 'Casarett and Doull's Toxicology'. McGraw-Hill, New York, 5th edn, 1996.

R.N.M. MacSween, and K. Whaley, eds, 'Muir's Textbook of Pathology'. Edward Arnold, 13th edn, London, 1992.

'Environmental Health Criteria No 199 – Principles and Methods for the Assessment of Nephrotoxicity associated with Exposure of Chemicals', International Programme on Chemical Safety, World Health Organization, Geneva, 1991.

CHAPTER 16

Neurotoxicity

JOHN H. DUFFUS

16.1 INTRODUCTION

Neurotoxicity is the term applied to a toxic effect on any aspect of the central or peripheral nervous system. Effects may be functional (behavioural or neurological abnormalities), neurochemical, biochemical, physiological, or morphological. It has been suggested that over one-third of chemicals may be neurotoxic in some sense.

16.2 THE NERVOUS SYSTEM

The central nervous system (CNS) is the main mass of nerve tissue lying between the effector and receptor organs, co-ordinating the nervous impulses between the receptors and effectors. The CNS is present in vertebrates as a dorsal tube which is modified anteriorly as the brain and posteriorly as the spinal cord; these organs are enclosed in the skull and backbone respectively. In higher organisms, the CNS takes on an additional activity in the form of memory which is the storage of past experiences.

The peripheral nervous system (Figure 16.1) connects with the CNS and is usually subdivided into two parts, somatic and autonomic. The somatic nerves supply voluntary skeletal muscles and skin sensory organs, and the autonomic nervous system is the part of the nervous system that controls the involuntary activities of the body. There are two main parts to the autonomic nervous system:

- the sympathetic nervous system, in which complexes of synapses form ganglia alongside the vertebrae. The preganglionic fibres from the CNS are short; these fibres are adrenergic, synaptic transmission being mediated by noradrenaline (norepinephrine).
- The parasympathetic nervous system, in which the ganglia are embedded in the wall of the effector. Thus, in this case, the preganglionic fibres are long and the postganglionic fibres are short; these fibres are cholinergic, synaptic transmission being mediated by acetylcholine.

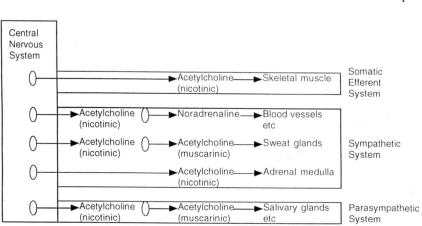

Figure 16.1 *The peripheral nervous system with nicotinic and muscarinic receptors*

The sympathetic and parasympathetic systems innervate the same end organs, but the effects produced by the two systems generally oppose one another.

The sympathetic system inhibits peristalsis, stimulates contraction of the sphincters of bladder and anus, inhibits bladder contraction, stimulates the heart pacemaker and speeds up the heart, stimulates arterial constriction, inhibits contraction of the bronchioles and permits their dilation, and inhibits contraction of iris muscle and permits dilation of the pupils.

The parasympathetic system stimulates peristalsis, inhibits contraction of the sphincters of bladder and anus, stimulates bladder contraction, inhibits the pacemaker and slows down the heart, inhibits arterial constriction and permits dilation, stimulates contraction of the bronchioles, and stimulates contraction of iris muscle and thus of the pupil.

The principal cells of nervous tissue are the neurones. Each neurone has a cytoplasm rich in granular endoplasmic reticulum (with granules of Nissl substance) carrying out active protein synthesis. Each neurone also has several short dendrites and an axon which can be very long. The axon contains no granular endoplasmic reticulum but it contains neurofilaments and microtubules which drive bidirectional movement of molecules and organelles. At the synaptic ending, the axon may endocytose neurotoxicants, viruses, and neurotransmitters and allow them to enter the neurone to be transported to the perikaryon (cell body).

In addition to the neurones, there are the glial cells which include macroglia (astrocytes and oligodendrocytes) and microglia. Astrocytes are star-shaped cells, the processes of which are often in close relation to endothelial cells and also with meningeal cells which cover the CNS; a trophic and axonal guiding role has been attributed to them. They may also contribute to the blood–brain barrier. Microglial cells show macrophage activity. The neuronal axon is often in very close relation to oligodendroglial cells in the CNS or to Schwann cells in the peripheral nervous system. Schwann cells are responsible for myelination.

The myelin sheath produced by them speeds up the transmission of electrical signals along peripheral nerves.

16.3 THE BLOOD–BRAIN BARRIER

The blood–brain barrier is formed by the endothelial cells of brain capillaries which differ from other endothelial cells. These endothelial cells are bound to each other by tight junctions which prevent the entry of proteins and other water-soluble substances of low molecular mass. Some areas of the nervous system lack a blood–brain barrier, the choroid plexus, neurohypophysis, eminentia media, pineal gland, area postrema, and the subfornical organ. In the brain, compartments filled with cerebrospinal fluid (CSF) communicate with the brain extracellular space. The CSF is secreted mainly by the choroid plexus. Among the cavities filled by the CSF are the ventricles, which are lined with ependymal cells (a variety of macroglia), and the subarachnoid space, which separates the pia mater from the arachnoid.

Sensory and autonomic ganglia are not protected by the blood–brain barrier and so their neurones are particularly exposed to neurotoxicants. Peripheral nerves have a special protective barrier isolating them called the perineurium. The perineurium is a connective tissue sheath; the inner layers of perineurial cells are connected by tight junctions which prevent free movement of fluid and solutes. There is also a blood–nerve barrier provided by capillary and endothelial cells bound by tight junctions.

16.4 SPECIAL FEATURES OF THE NERVOUS SYSTEM

Cells of the nervous system have high energy requirements: they have glucose as their only substrate and are unable to use ketones. Thus, they are highly dependent on a good blood supply of both glucose and oxygen for oxidative phosphorylation. Although many of the cells have blood–brain barrier protection, the nervous system extends throughout the body with a complex geometry and so there are many possible vulnerable areas. Further, the way in which the nerve impulse is transmitted chemically across extracellular space is particularly at risk from interference by toxicants.

Neurones are unusual in that they require an environment rich in lipids to be maintained in order to function properly. This may be one of the reasons that neurones are unable to regenerate. The inability of neurones to regenerate makes it important to prevent any neuronal death, as excessive loss of neurones over a lifetime may lead to dementia in old age.

16.5 TOXIC INJURY TO THE NERVOUS SYSTEM

The main toxic injuries to which the nervous system is liable are neuronopathy, axonopathy, myelinopathy, carcinogenesis, developmental injury, and complex damage leading to behavioural abnormalities.

16.5.1 Neuropathy

There are various kinds of neuropathy. These are listed below with some of the causative agents.

- Neuronal chromatolysis is the disappearance of Nissl substance. It may be caused by methylmercury acting on dorsal root ganglion neurons, alkyl lead acting on brain stem neurones, and podophyllotoxin acting on spinal ganglion neurones.
- Neuronal atrophy and degeneration is caused by acrylamide (acute toxic peripheral neuropathy) and by trimethyltin.
- Neuronal swelling is caused by trimethyltin in the brain stem neurones and in the anterior horn cells of the spinal cord.
- Neurofibrillary degeneration is caused by eating cycad nuts (Parkinsonism-dementia of Guam), aluminium (in the cerebral cortex and hippocampus neurones), organolead, and podophyllotoxin.
- Dendropathy, damage to the dendritic processes, is caused by alcohol, inorganic lead, and mercury.

16.5.2 Axonopathy

Axonopathy is caused by β-iminodipropionitrile, n-hexane, methyl n-butyl ketone (all proximal); acrylamide (distal); and generally by carbon disulfide, isoniazid, inorganic lead, leptophos, polychlorinated and brominated biphenyls, taxol, trichloroethylene and triorthocresyl phosphate. It should be noted that the symptoms, which could result from a single dose of triorthocresyl phosphate, may not be apparent for 10–14 days after exposure.

16.5.3 Myelinopathy

This is the destruction of myelin sheaths around the axons, and is caused by triethyltin, inorganic lead, acetylethyltetramethyl tetralin (AETT), cuprizone, cyanate, dichloroacetate, hexachlorophene, and isoniazid.

16.5.4 Toxic Degeneration of the Synapse

The following toxic actions may occur at the synapse:

- Inhibition of the action potential
- Blocking of ion channels, altering electrical transmission and intracellular signalling
- Interference with synthesis of a neurotransmitter
- Interference with storage, release, or uptake of neurotransmitter
- Binding of neurotransmitter to a target cell receptor
- Enzymic breakdown of excess or unused transmitter.

Among the substances that produce such effects are anatoxins, batrachotoxin, botulinum toxin, cocaine, α-dendrotoxin, inorganic lead, nicotine, organophosphorus insecticides, pyrethroid insecticides, saxitoxin, scorpion toxins, tetanus toxin, and tetrodotoxin

16.6   MECHANISMS OF PRODUCTION OF NEURONAL LESIONS

In general, lipid-soluble compounds (as reflected by their octanol/water partition coefficient) enter the brain easily. This is why methylmercury, for example, is more neurotoxic than the mercuric ion.

Mechanisms of chemical damage to the nervous system can be of any kind known to toxicology and so it is possible to list only some of the more common examples.

- Methylmercury. Methylmercury causes disapearance of Nissl substance through an action on protein synthesis. It also acts on −SH groups in membranes, altering permeability and transport systems. In the cerebral cortex this causes degeneration and necrosis of neurones, especially those dealing with vision.
- Carbon monoxide. Carbon monoxide binds reversibly (but more avidly than oxygen) to the iron in haemoglobin, myoglobin, and the cytochromes, so that oxygen-carrying capacity is reduced. Oxygen reaches the tissues less easily and its use in oxidative phosphorylation is inhibited. This inhibits neurone metabolism causing peripheral neuropathy, and by depriving the brain of oxygen causes loss of consciousness.
- 2,5-hexanedione, carbon disulfide, and acrylamide. These inhibit some glycolytic enzymes directly by reacting with proteins to form pyrrole adducts.
- Arsenic. Arsenic binds to lipoic acid which is involved in pyruvate decarboxylation. This causes symptoms similar to thiamine deficiency.
- Thallium. Thallium binds to mitochondrial membranes and causes similar symptoms to arsenic, resembling thiamine deficiency.
- Aluminium. Aluminium causes abnormal accumulation of neurofilaments in the perinuclear region of neuronal cell processes so that other cell organelles are displaced peripherally. Aluminium intoxication has been observed in patients undergoing renal dialysis with aluminium-containing dialysates and results in progressive dementia.
- Tetrodotoxin. Tetrodotoxin blocks sodium channels in the nodes of Ranvier causing paralysis by inhibition of peripheral nerve function.
- Botulinum toxin. Botulinum toxin blocks the release of acetylcholine at the synaptic ending of the neurone.
- Black widow spider venom. This causes continuous release of acetylcholine.
- Snake venom bungarotoxin. Snake venom bungarotoxin binds to acetycholine postsynaptic receptors, blocking neurotransmission and causing paralysis.
- Organophosphorus compounds. Distal axonopathies occur in both the

CNS and the peripheral nervous system caused by reduction in the number of microtubules in the cytoskeleton. This was first observed in neuropathies due to chronic exposure to organophosphorus compounds. Organophosphorus compounds cause acute damage to the nervous system by phosphorylating the neurone membrane-bound protein, acetylcholine esterase, essential for the breakdown of acetylcholine and therefore for the maintenance of normal neurotransmission.

- Diptheria toxin. Diphtheria toxin causes lesions in Schwann cells and in oligodendrocytes by inhibiting the synthesis of the proteolipid of myelin and of the basic protein of the Schwann cell, thus slowing conduction in the affected axons.
- Other toxic effects. Triethyltin also affects the myelin sheath, causing widespread oedema in the white matter of the brain. Hexachlorophene neurotoxicity is similar in effect.

Care has to be taken in interpreting these effects if they are seen in experimental animals as they may be species-specific. For example, acetylethyltetramethyltetralin is used in scent and does not normally cause toxicity in humans. However, it causes vacuolation of myelin in experimental animals.

Some toxic substances show a degree of specificity for the part of the nervous system on which they act. For example, tetanus toxin specifically binds to the neuronal surface of the spinal cord motor neurone and may be used to distinguish neurons from non-neuronal cells. It inhibits the release of γ-aminobutyric acid and glycine, causing muscle spasms and rigidity.

Some toxicants attack the blood–brain barrier itself, *e.g.* mercury. The critical organ for inorganic mercury is the kidney but the main target for organic mercury is the nervous system and part of its effect is due to damage to the blood–brain barrier.

The mechanism of production of nervous system tumours is presumably very similar to that of tumours of other organs (see Chapter 9). Some of the chemicals known to induce such tumours in rats are *N*-methyl-*N*-nitrosourea, *N*-ethyl-*N*-nitrosourea, azoxyethane, 1,2-diethylhydrazine, 1-phenyl-3,3-diethyltriazine, acrylonitrile, and ethylene oxide. Whether these substances have similar effects in humans is a question still to be answered by epidemiological studies.

16.7   BIBLIOGRAPHY

B. Ballantyne, T. Marrs and P. Turner, eds. 'General and Applied Toxicology' (2 volumes). Macmillan, London, 1993.

C.D. Klaassen, ed. 'Casarett and Doull's Toxicology'. McGraw-Hill, New York, 5th edn., 1996.

R.N.M. MacSween and K. Whaley, eds. 'Muir's Textbook of Pathology'. Edward Arnold, London, 13th edn., 1992.

CHAPTER 17

Behavioural Toxicology

GERHARD WINNEKE

17.1 INTRODUCTION

The study of behaviour and its reversible or irreversible modification by chemicals or therapeutic agents in toxic doses is a relatively new approach within neurotoxicity. Generally, neurotoxicology draws upon a broad spectrum of neuroscience methods, such as neuroanatomy, neurochemistry, neurophysiology, neurogenetics, or neuropharmacology to study adverse effects of chemicals on the structure and function of the mature or developing nervous system, and neurobehaviour is part of this spectrum. Adding the prefix neuro to behaviour has been proposed to emphasize this link with the other neurosciences.

From a historical perspective behavioural toxicology has evolved from empirical psychology, which has also been defined as the objective study of behaviour, and from strong methodological roots of animal studies within experimental and comparative psychology. Despite its closeness to these other neuroscience approaches, behavioural toxicology also has clear, distinctive features. Behavioural changes are dependent on underlying biochemical or physiological changes at various morphological sites within the body, and behavioural analysis deals with the integrated output of these bodily changes, which may not even be restricted to the nervous system.

There is a tendency to use the term behavioural toxicology for animal models and the term neuropsychological toxicology for the behavioural consequences of human exposure settings. This distinction will not be made in this review. Instead, selected human and animal models and typical studies will be treated under the same heading of behavioural toxicology. It has also become customary to discuss studies on postnatal behavioural deficit following pre- or perinatal exposure to chemicals under the heading of behavioural teratology. Again, although this subspeciality within the general field of behavioural toxicology is important from a general neurobiology point of view, the distinction between adverse effects of chemicals on the adult *vs.* the developing brain will not be considered here, because it is of only minor importance for the selection of behavioural models.

17.2 ANIMAL APPROACH TO BEHAVIOURAL TOXICOLOGY

It is beyond the scope of this review to describe in any detail the principles underlying behavioural pharmacology and toxicology. This has been done elsewhere by Tilson and Harry. Much of the terminology and methodology in this field has been developed in the behaviourist tradition of experimental and comparative psychology. Within this is traditional behaviour, defined as the movement of an organism or its parts in a temporo/spatial context, which can be either unconditioned (unlearned) or conditioned (learned). Unconditioned behaviour may either be respondent, *i.e.* elicited by known, observable stimuli, or operant, *i.e.* emitted in the absence of any observable eliciting stimulus. Examples for the former are reflexes or goal-directed movements, examples of the latter are muscular or spontaneous locomotor activities. Conditioned or learned behaviour includes classical or type S conditioning in the Pavlovian sense, whereby new stimulus response associations are formed by repeated pairing of an unconditioned behaviour with a neutral stimulus to become a conditioned stimulus, and instrumental conditioning, whereby behavioural changes are brought about by the consequences of a behavioural response, namely reward, punishment, or negative reinforcement.

Animal models used in behavioural toxicology are available in each of the four classes of behaviour mentioned above, namely unconditioned/respondent, unconditioned/emitted, classically conditioned, and instrumentally conditioned. For each of these classes, selected typical models will be described briefly and explained by illustrative examples. Chemical-induced deficits of motor and sensory functions are typically assessed by means of the unconditioned/respondent and unconditioned/emitted models, whereas the classically conditioned and instrumentally conditioned models may be used for the assessment of chemical-induced cognitive deficits, *i.e.* learning and memory dysfunction.

The distinction between learning and memory is somewhat artificial and more a matter of emphasis rather than of substance. If learning is defined as an adaptation to environmental change, memory means the retention of such adaptation over some period of time. Learning requires intact memory and *vice versa*; memory dysfunction cannot be tested in the absence of a well-learned task. It follows that, in practice, neither function can be studied in isolation, although it is also true that in some behavioural procedures the emphasis is more on learning or acquisition of behavioural adaptations, whereas in others retention is more prominent. This will be considered in the subsequent sections.

Although animal paradigms for use in behavioural toxicology have been adapted for different species, including pigeons, rodents, and non-human primates, the animal part of this brief overview will, with few exceptions, concentrate on rodent paradigms. The preferred animal model in behavioural toxicology is the rat. It should also be stressed that much detail of methodology cannot be given here. The interested reader should consult a more practical text such as Sahgal (1993) for more detailed information.

17.2.1 Unconditioned Respondent Behaviour

Examples of unconditioned respondent behaviour are the acoustic startle reflex (ASR) and its amplification, the prepulse inhibition technique or reflex modulation (RM). These techniques are widely used in behavioural toxicology for testing the excitability of the central nervous system (CNS) and/or muscle weakness, or for the measurement of hearing thresholds in rodents (RM). In ASR testing, the animal is placed on a strain gauge-coupled platform in a sound attenuated chamber and a short burst of white noise of 120 dB sound presssure level (SPL) is presented. The amplitude of the accompanying motor reflex is measured. Since this measurement is ambiguous and may contain excitability, motor, as well as sensory components, more specific information about auditory functioning can be obtained in RM. In this procedure the eliciting stimulus of 120 dB SPL (S_2) is preceded by brief prepulse stimuli of different intensities from faint to loud (S_1) in random order, and the amplitude reduction for the S_2 response associated with the prepulse stimulus is recorded as a function of the S_1 intensity. This allows for auditory threshold testing in rats and a complete audiogram may be established in about 3 h.

An illustrative example is the assessment of auditory deficits following 3,3'-iminodipropionitrile (IDPN) administration in the rat. IDPN (0, 100, 150, and 200 mg kg^{-1} day^{-1}; $n = 9/10$) was given i.p. in saline on three consecutive days. Animals were tested 9 weeks post-exposure for ASR amplitude and RM as described above. The results are shown in Figure 17.1.

There is a dose-dependent decrease of ASR amplitude, whereas auditory thresholds at 5 and 40 kHz are only affected at the highest dose. This suggests that both effects are unrelated, and that additional mechanisms, such as damage of the primary neuronal pathway for eliciting the ASR, may be involved in the ASR effect. This also proved to be different from the RM effect in its time course (not shown). Ototoxicity is irreversible as shown in a second exposure group tested at 126 days post-exposure. The hindlimb grip strength was not affected.

By combining ASR and RM in a similar study it was shown that exposure to triethyltin produces ASR reduction in the absence of ototoxicity, the likely interpretation being in terms of muscle weakness.

Other variables in this behavioural category include developmental milestones, such as placing, homing, the righting reflex, negative geotaxis, and different measures of forelimb or hindlimb grip strength or of motor co-ordination.

17.2.2 Unconditioned Emitted Behaviour

A typical example in this behavioural category is motor activity (MA). This is often regarded as an apical test of neurotoxicity, because it is influenced by sensory, motor, and even associative processes in the nervous system. In this capacity MA tests are typical components of screening batteries.

Many different devices are available for motor activity testing, ranging from simple open-field arenas to complex photocell-controlled alleyways. The functional comparability of different devices has always been a matter of concern. In

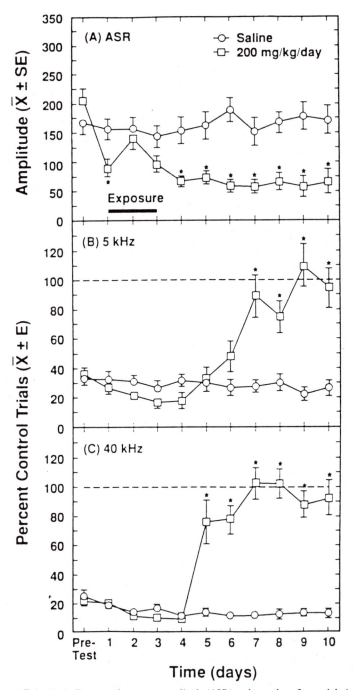

Figure 17.1 *(A) Auditory startle response amplitude (ASR) and prepulse reflex modulations to (B) 5 kHz and (C) 40 kHz following three consecutive iminodiproprionitrile (IDPN) injections into rats*

(This figure has been reproduced by kind permission of the publisher from K.M. Crofton, and T. Knight, Auditory deficits and motor dysfunction following iminodipropionitrile administration in rats, *Neurotox. Teratol.*, 1991, **13**, 575–581)

order to clarify the issue an interlaboratory comparison study with a group of stimulants, depressants, and other chemicals as well as a number of different photocell-to-actometer devices in circular, quadratic, or figure-8 form was conducted. The qualitative similarity of outcome across laboratories and devices was reassuring. Except for cypermethrin, the results were consistent, as was the activity decrease over time (habituation). It was concluded that the results of this comparative study were consistent for different chemicals and laboratories, although no explicit effort was made to standardize for age, strain, or testing protocol.

Other variables in this behavioural category include measurements of emotion in the open-field test (defecation, urination) or through ultrasonic vocalizations, sleep–wakefulness patterns, and consummatory as well as reproductive activities.

17.2.3 Classically Conditioned (Learned) Behaviour

This type of associative learning or type S conditioning, which was first studied and described by Pavlov, is based on the principle of temporal contiguity. An unconditioned stimulus (US), *e.g.* food in the mouth, elicits a reflex-type unconditioned response, *e.g.* salivation, whereas a neutral stimulus (NS), *e.g.* sound, is ineffective in this respect. By systematically pairing NS and US, the NS gradually acquires the ability to elicit the response, which was initially associated with the US alone, and is therefore turned into a conditioned stimulus (CS). The response to the CS is now termed a conditioned response (CR).

Paradigms based on principles of classical conditioning are not widely used in behavioural toxicology. One remarkable exception is the classical conditioning of the nictitating-membrane extension reflex in the rabbit. The US in this model was paraorbital shock and the CS a tone. Following systemic injection of aluminium lactate, acquisition of the conditioned response was impaired at the two highest dose levels (200, 400 μmol kg^{-1} body weight).

Another example of conditioned behaviour is conditioned taste aversion. A classical approach is the pairing of lithium chloride, an emetic agent, with saccharin-containing water. The typical outcome is that the subsequent intake of sweetened water is reduced relative to pre-conditioning intake. Applications in behavioural toxicology include 2,4,5-trichlorophenoxyacetic acid.

17.2.4 Instrumentally Conditioned (Learned) Behaviour

The principle of instrumental conditioning or instrumental learning, as formulated by Thorndike in his Law of Effect, is that behaviour is controlled by its consequences: 'Of several responses made to the same situation, those which are accompanied or closely followed by satisfaction to the animal will ... be more firmly connected with the situation, so that when it recurs, they will be more likely to recur; those which are accompanied or closely followed by discomfort to the animal will ... have their connections with that situation weakened, so that when it recurs, they will be less likely to occur. The greater the satisfaction or discomfort, the greater the strengthening or weakening of the bond'.

Most of the paradigms used in behavioural toxicology for the assessment of cognitive deficit, *i.e.* disruption of learning and memory, belong to this class of instrumental learning. A variety of different procedures is available and the choice among these depends on the question being asked. Although, as has already been pointed out, the distinction between memory or retention on the one hand, and learning or acquisition on the other, is somewhat arbitrary, different models may nevertheless be classified as primarily memory-directed or learning-directed.

17.3   RETENTION-DIRECTED MODELS

17.3.1 Passive Avoidance Conditioning

This is a simple retention test (long-term memory) based on one-trial learning. In the step through model the animal is shocked as soon as it enters a darker compartment from a brightly lit one through a guillotine door connected to a runway. In the step down procedure the animal receives a foot shock as soon as it steps down from a pedestal on a grid floor. Animals are retested at varying intervals afterwards, *e.g.* 5 min, 3 h, 24 h, and latencies to re-enter or to step down are measured. Good retention is characterized by prolonged latencies upon retesting.

Using this paradigm in trimethyltin-treated rats, a dose-dependent reduction or re-entering latencies has been shown 24 h after initial training. Although passive avoidance tests are quick and economical, performance in this task is influenced by many factors, *e.g.* pain sensitivity, reactivity, motor activity. The interpretation of outcome may therefore require additional information from other behavioural models.

17.3.2 Delayed Alternation

In this task the animal is required to alternate between two different responses to receive reinforcement. A delay element between choices is introduced to cover the short-term memory aspect. Testing may be done in Skinner boxes with two levers or in t-mazes (right/left turn). A marked performance deficit in a delayed alternation task has been shown in 5–6 year old rhesus monkeys exposed to inorganic lead during the first year of life. Disruption of delayed t-maze spatial alternation has been found in female rats 90 days postnatally after gestational and lactational exposure to specific ortho-substituted polychlorinated biphenyl congeners.

17.3.3 Delayed Matching or Non-Matching to Sample

This test is the animal equivalent of recognition memory in man. Although initially developed for use in monkeys, adaptations for rats exist. In both tasks the animal is first presented with a sample stimulus which is rewarded and then, after some delay, it is given a choice between the sample and a novel stimulus. In the matching variant the animal is rewarded for choosing the original, and in the non-matching model for selecting the novel stimulus. Since the information to be remembered in each trial is independent from the next one, both tasks are mainly tests of working (short-term) memory rather than of reference (long-term) memory. The delayed non-matching to sample oddity problem is the easier task because it takes advantage of the animal's spontaneous novelty preference. The delay element emphasizes the short-term memory component. It is typically done in y-mazes with visual stimuli placed at the end of each of the three arms, and equipped with guillotine doors to produce varied intervals of delay. Applications are primarily in behavioural pharmacology.

17.3.4 Radial Arm Maze

The radial arm maze (RAM) was originally developed by Olton as an octagonal arena from which eight arms radiate outwards. The RAM has become popular as a behavioural paradigm to specifically tap spatial working and reference memory. It is easily acquired by normal rats and a variety of other species as well, can be automated, and is highly sensitive to hippocampal lesions. Detailed protocols are described by Rawlins and Deacon (Sahgal, 1993). Food-deprived rats are allowed to choose freely from among the eight arms, but only one food pellet is available at the end of each. The rat is free to choose arms continuously until all novel arms have been entered and food taken. Generally rats soon learn to avoid re-entering arms that have already been visited.

The RAM can be used to differentiate between working or short-term memory and reference or long-term memory in the following manner. Always baiting the same four arms and leaving the remaining unbaited results in the learning of the rat to reliably avoid the set of unbaited arms (reference memory), as well as the baited arms already successfully visited during that particular session (working memory).

Early developmental lead exposure was found to impair acquisition and retention of RAM performance. Reacquisition of RAM performance was impaired in rats treated with trimethyltin.

17.3.5 Morris Watermaze

This test is an alternative to the RAM in that it is considered a model for the assessment of spatial learning and retention, and has also been shown to be sensitive to hippocampal dysfunction. Typical protocols are given in Stewart and Morris (Sahgal, 1993). It is a swim test and the acquisition element is very simple. Animals must learn the spatial position of a hidden platform from varying starting positions. If the learning criterion is reached, retention is tested after removal of the platform as a percentage of time spent in the correct quadrant.

Specific pharmacological interference with normal hippocampal functions, such as application of N-methyl-D-aspartate receptor antagonist AP5, has been found to disrupt learning and retention in the Morris watermaze. Use in behavioural toxicology is as yet limited.

17.4   LEARNING-DIRECTED MODELS

17.4.1 Active Avoidance Learning

Active avoidance learning is a frequently used paradigm in behavioural toxicology. It has a stronger emphasis on acquisition than on retention and creates active avoidance conditioning, either by one-way step through or pole jump avoidance or by active two-way avoidance. In one-way or pole jump avoidance, one-directional movements result in foot shock avoidance. Active pole jump avoidance conditioning with warning stimuli varying in pitch, has been used to demonstrate high-frequency hearing loss in rats exposed to toluene at weaning or in adult age. (See Figure 17.2.)

Bidirectional movement or shuttling between two compartments is necessary for avoiding the signalled foot shock in two-way active avoidance conditioning. This is a frequently used model in behavioural toxicology. Two-way active avoidance conditioning in behavioural toxicology is used to assess components of learning and memory as affected by neurotoxic chemicals. In an example of this, Wistar rats were given a commercial mixture of polychlorinated biphenyls (PCBs) such as Chlophen A30 in the diet for 60 days before mating. The results of testing are shown in Figure 17.3. Two-way active avoidance conditioning was impaired in offspring with in-utero exposure, but not in those with only postnatal exposure. It should be pointed out that pronounced group differences were seen on the second day of testing only, which is suggestive of a retention component in the observed PCB-related deficit. At the time of testing PCB congeners were no longer detectable in the brains of the offspring. These findings support suggestive evidence from epidemiological work about the particular importance of prenatal exposure for postnatal neurodevelopmental deficit.

Other examples include deficit of active avoidance learning after neonatal exposure to chlordecone, and prenatal exposure to low levels of carbon monoxide.

Caution is necessary when interpreting results from two-way active avoidance conditioning in terms of cognitive factors. Non-associative factors, such as changes in motor activity or pain sensitivity, may contribute to the observed outcome. Motor activity is typically controlled by counting inter-trial activity within the ongoing conditioning procedure. Alteration of pain sensitivity due to the experimental manipulation can be controlled by measuring pain thresholds using standard methods such as the hot plate test or the tail flinch response. The memory or retention aspect in this task can be emphasized when comparing acquisition curves in consecutive sessions (see above).

17.4.2 Repeated Chain Acquisition

Once a task has been learned by certain criteria, performance changes on this task can subsequently be tested. However, it is unfortunate if the time course of agent-induced learning impairment is the focus of interest. Repeated chain acquisition (RCA) is an elegant method of overcoming this difficulty, because new

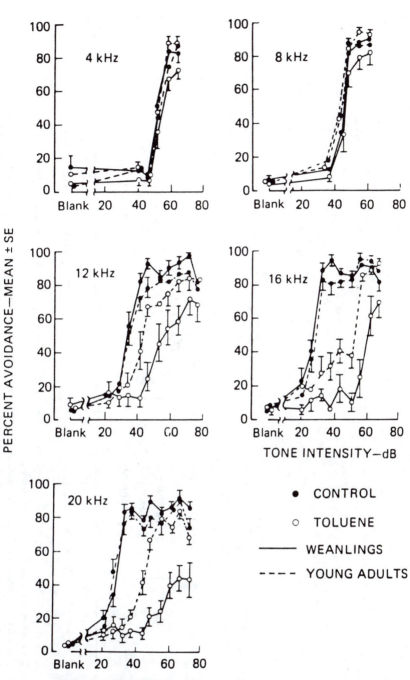

Figure 17.2 *Tone intensity response functions in rats first exposed to toluene as weanlings or as young adults in a pole jump, active avoidance model; effects are not apparent at low-frequency sounds below 12 kHz*

(This figure has been reproduced by kind permission of the publishers from G.T. Pryor, J. Dickinson, E. Feeney, and C.S. Rebert, Hearing loss in rats first exposed to toluene as weanlings or as young adults, *Neurobehav. Toxicol. Teratol.*, 1984, **6**, 111–119)

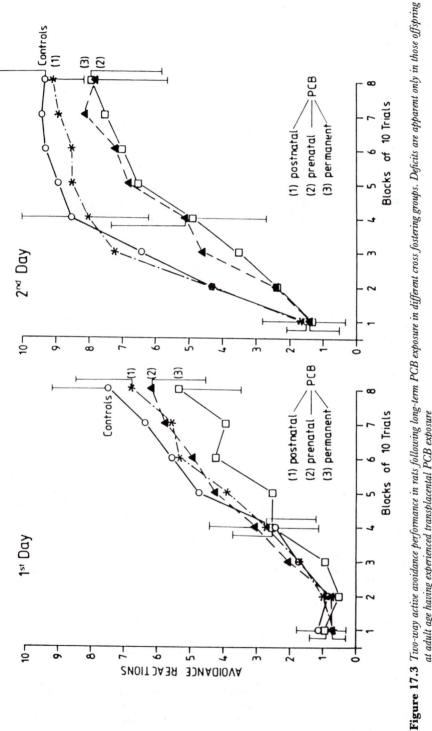

Figure 17.3 *Two-way active avoidance performance in rats following long-term PCB exposure in different cross fostering groups. Deficits are apparent only in those offspring at adult age having experienced transplacental PCB exposure*

(This figure has been reproduced by kind permission of the publishers from H. Lilienthal and G. Winneke, Sensitive periods for behavioural toxicity of polychlorinated biphenyls: Determination by cross fostering in rats, *Fundam. Appl. Toxicol.*, 1991, **17**, 368–375)

learning is required on each in a series of repeated learning sessions. If for example three levers are available, the animal may be required to press these levers in a different sequence on each successive session, or to respond to a different sequence of coloured cue lights on each session in order to be rewarded.

In the rat, RCA learning was found to be disrupted following a single injection of trimethyltin; impairment was found to last for at least 5 weeks post-injection.

17.4.3 Schedule-controlled Operant Behaviour

This is based on variants of reinforcement schedules which have been applied successfully in behavioural toxicology (Laties and Wood, 1986). In these behavioural models, pressing a lever for reward in a Skinner box is modified by the specific behaviour reinforcement contingencies. Thus, a food pellet may be given after a fixed or variable time interval has elapsed since the last response or else after a fixed or variable number of responses has been emitted. Each of these contingencies produces characteristic response patterns, the modifications of which have been studied following exposure to neurotoxic chemicals, such as inorganic lead and PCBs.

Schedule-controlled operant behaviour in the Skinner box is typically used in drug discrimination learning. This is a powerful behavioural tool to test the involvement of specific transmitters in chemically induced neurotoxicity. Receptor agonists and antagonists or saline are injected and conclusions on the involvement of specific receptor families are derived from subsequent differential responses to two levers, namely the 'drug' lever and the 'saline' lever.

17.5 SUMMARY OF ANIMAL MODELS

An impressive variety of animal models is available to cover motor, sensory, and cognitive functions, only a selection of which has been discussed here. Applications in the area of neurotoxicity have shown their value in screening for neurotoxic potential of chemicals, in helping to establish cause and effect as well as dose–response reactions, contributing to an understanding of the neural mechanisms underlying neurotoxicity, and in identifying critical periods of exposure.

It should also be pointed out, however, that this developing field is still characterized by a certain eclecticism rather than by an established set of unifying principles capable of integrating the findings produced by a large and varied set of paradigms. It should also be emphasized that most animal models are mere structural analogues of neuropsychological functions in man, and that efforts need to be intensified to develop behavioural models that allow for a better functional extrapolation to effects observed at the human level.

17.6 THE HUMAN APPROACH TO BEHAVIOURAL TOXICOLOGY

Experimental as well as field studies on behavioural effects of occupational and environmental exposure need to be considered here.

Table 17.1 *Selection of domains, behavioural tests, and functions typically covered in experimental human exposure studies*

Domain	Test	Function
Psychomotor performance	Simple reaction time	Visuomotor speed
	Choice reaction time	Visuomotor speed
	Tapping	Motor speed
	Pursuit tracking	Co-ordination
	Steadiness, Aiming	Tremor, co-ordination
	Digit span	Attention span
Attention	Mackworth clock	Sustained attention
	Auditory vigilance	Sustained attention
Visual perception	Tachistoscopic vision	Perceptual speed
	Critical flicker fusion	Temporal resolution
Cognition	Benton test	Visual memory
	Paired associate	Verbal memory
	Learning and recall	Verbal memory
Affect	Mood scale	Mood

17.6.1 Experimental Human Exposure

Experimental exposure studies in human subjects are conducted mainly to identify the presence or absence of acute neurobehavioural effects at low levels of exposure that may serve as indicators of subclinical effects, or to check for an impairment of human performance capabilities by short-term exposures to neurotoxic chemicals, which may increase the risk of unsafe job performance.

Many human experimental studies have been carried out using a variety of tests, but these have largely concentrated on the functional domains and associated tests shown in Table 17.1. Although not exhaustive, this list can be regarded as representative of the type of tests and functions covered in most of the human chamber studies up to about 1986. More recent studies have used computerized tests such as the Neurobehavioural Evaluation System (see below and Table 17.2.) which, from a functional point of view, does not differ substantially from the tests listed in Table 17.1.

Among the chemicals studied, organic solvents have received particular attention, namely chlorinated hydrocarbons (methylene chloride, trichloroethylene, tetrachloroethene, 1,1,1,-trichloroethane), aromatic hydrocarbons (toluene, styrene, xylene), ketones (acetone, methyl ethyl ketone), carbon monoxide, and anaesthetics. Exposure levels have typically been at or below threshold limit values (TLVs) for the individual compounds.

Although results differ somewhat both between and within the same chemicals, depending on the level and/or the duration of exposure, a typical finding for a group of solvents such as the halogenated hydrocarbons, is impaired vigilance

PSYCHOMOTOR   PERFORMANCE

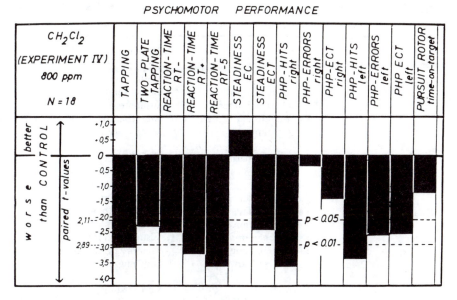

Figure 17.4 *Impairment of psychomotor performance in humans after* 4 h *of exposure to* 800 ppm *methylene chloride vapour. Values on the ordinate are descriptive* t-*values relative to control condition within subjects*
(This figure has been reproduced by kind permission of the publishers from G. Winneke, 'Behavioural effects of methylene chloride and carbon monoxide as assessed by sensory and psychomotor performance', in 'Behavioural Toxicology. Detection of Occupational Hazards', eds. C. Xintaras, B.L. Johnson, and I. de Groot. US Department of Health, Education and Welfare, 1974, pp. 130–344)

and generalized slowing of perceptual and psychomotor functions suggestive of CNS depression or a prenarcotic state. This is exemplified by the spectrum of psychomotor impairment after 4 h of single blind exposure to 800 ppm of methylene chloride as shown in Figure 17.4. In addition to psychomotor impairment, vigilance decrement and elevated flicker fusion thresholds were reduced at concentrations as low as 300 ppm.

Experimental studies and effects such as these have in some instances contributed to the confirmation or modification of existing TLVs for occupational exposure, although there is as yet no conclusive evidence that acute neurotoxic exposure has indeed contributed to accident-prone workplace performance.

17.7   FIELD STUDIES: OCCUPATIONAL EXPOSURE

A variety of different behavioural tests and batteries of tests have been, and are still being used in studying neurotoxic effects associated with workplace exposure to chemicals. Typical domains, tests, and functions used in many occupational field studies are included in the computerized Neurobehavioural Evaluation System and are illustrated in Table 17.2. Other computerized collections of tests

Table 17.2. *Tests and functional domains covered in the computerized Neurobehavioural Evaluation System*

Domain	Test	Function
Psychomotor performance	Symbol digit	Coding speed
	Hand–eye co-ordination	Co-ordination
	Simple reaction time	Visuomotor speed
	Continuous performance test	Attention span
	Finger tapping	Motor speed
Perceptual ability	Pattern comparison	Visual perception
Memory and learning	Digit span	Short-term memory/attention
	Paired associate learning	Visual learning
	Paired associate recall	Intermediate memory
	Visual retention	Visual memory
	Pattern memory	Visual memory
	Memory scanning	Memory processing
	Serial digit learning	Learning/memory
Cognitive	Vocabulary	Verbal ability
	Horizontal addition	Calculation
	Switching attention	Mental flexibility
Affect	Mood test	Mood

include the Swedish Performance Evaluation System and the World Health Organisation (WHO) Neurobehavioural Core Test Battery.

The groups of chemicals of primary concern in behavioural toxicity include solvents, metals and pesticides. Most behavioural studies, however, have concentrated on solvents and metals. Some examples are given to illustrate the main approaches and typical findings.

17.7.1 Organic Solvents

Solvents represent a large, chemically heterogenous group of compounds which are liquid within the range of 0–250 °C, and are widely used for extracting, solution, or suspension of water-insoluble materials. They may be grouped into a few broad classes, namely aliphatic, aromatic, halogenated hydrocarbons, alcohols, ketones, esters, and mixtures. Due to the wide variety of applications, occupational exposure is frequent. Because of their lipophilic nature, the nervous system is the primary target for inhaled solvents. Their narcotic action is the predominant biological effect in the CNS. Functional, as well as structural, effects ranging from neurophysiological changes to severe polyneuropathies have been reported to occur in the peripheral nervous system. Comprehensive reviews covering relevant aspects of chemistry, exposure, and biomedical effects such as Riihimäki and Ulfvarson (1986) should be consulted for more detailed information.

Much of the CNS neurotoxicity of organic solvents may be explained in terms of narcotic action. Prenarcotic reversible effects, such as psychomotor slowing or vigilance decrement, have been documented in experimental human exposure studies (see above). As yet it is not clear, however, whether repeated prenarcotic exposure over the years may eventually result in irreversible brain damage, but it has been shown that for some compounds, such as trichloroethylene, styrene, and carbon disulfide, as well as solvent mixtures, chronic low-level exposure is associated with perceptual, cognitive, and motor retardation which, from the very design of the different studies, could not be explained as acute reversible effects. In summarizing several such studies the following conclusion was drawn by Gamberale '... that the measurement of behavioural performance has been demonstrated to possess more general applicability in human studies than other methods'.

17.7.2 Metals

Lead, mercury, and manganese have received particular attention in occupational exposure studies using behavioural endpoints. Both cross-sectional and prospective studies are used in considering behavioural effects of neurotoxic metals in humans. In cross-sectional studies, the preferred approach in occupational exposure studies, behavioural functions are measured in exposed subjects at one point in time and typically related to concurrent exposure, either relative to unexposed, but otherwise comparable subjects, or relative to one continuous exposure variable using regression models. In prospective studies, associations are established between changes in behavioural functions relative to the natural exposure history. This approach offers an opportunity to identify critical periods of exposure and to clarify if the observed effects are reversible or not. Furthermore, problems of causality are more easily handled in prospective studies.

17.7.2.1 Lead. Lead is probably the best studied toxic element. A comprehensive review covering the chemical, environmental, and biomedical aspects has been published by the International Programme on Chemical Safety of the World Health Organization (1995). This metal occurs in both organic and inorganic forms, namely as lead salts of widely different water solubilities. Although the organometallic lead species, due to their lipid solubility, are highly neurotoxic in acute exposure situations, chronic low-level exposure to inorganic lead constitutes the more serious occupational and public health risk.

A broad spectrum of biomedical effects has been found to be associated with lead, the more critical of which are related to haem biosynthesis, erythropoiesis, and nervous system function. The latter has received considerable attention in terms of occupational health and environmental exposure settings using behavioural performance as the main outcome variable.

Cognitive as well as sensorimotor functions have often been studied in occupationally exposed subjects. Cognitive functions are typically covered by means of results from standard intelligence tests, such as the Wechsler Adult Intelligence Scale. Tests for the assessment of sensorimotor functions include different reaction-time models and tasks of sensory motor speed and co-ordination, as well as tasks of perceptual speed and sustained attention.

Several studies have been conducted which indicate that behavioural deficit does not occur at blood lead levels below 2.0–2.5 μmol l^{-1}. In one study (Stollery *et al.*, 1989) tasks were selected on theoretical considerations based on models of cognitive information processing rather than purely on clinical grounds. Ninety one workers from battery and printing industries with low (1.0 μmol l^{-1}), medium (1.0–2.0 μmol l^{-1}), and high (2.0–4.0 μmol l^{-1}) blood lead levels were given computer-based tests for the assessment of serial sensory motor reaction time, visual memory, attention, verbal reasoning, and spatial processing.

After allowing for several non-lead variables such as years of exposure, years of schooling, alcohol consumption, and work demand, a significant exposure-related behavioural deficit was found for decision and movement times, and for decision gap rates of the serial reaction-time task, but not for cognitive performance. Performance was affected only when the blood lead level was greater than 2.0 μmol l^{-1}. The conclusion was drawn that the sensorimotor rather than the cognitive components of the behavioural tasks provided the most sensitive indicators of the effects of long-term occupational lead exposure, and that the effect on gap rates could be interpreted in terms of lead-related attentional difficulties.

17.7.2.2 Mercury.   Like lead, mercury belongs to those metals known to, and used by, man since ancient times. For a more detailed coverage of the chemical, environmental, and biomedical aspects of mercury toxicity the reader is referred to Berlin (1986). Mercury has three oxidative states, namely metallic mercury, the mercurous ion Hg(I) and the mercuric ion Hg(II). Each has unique characteristics on target organ toxicity. Exposure to metallic mercury vapours is limited to occupational exposure and that due to amalgam fillings. The ability of mercury to form stable organic compounds with alkyl groups, which are usually referred to under the common heading of methylmercury, are of public health significance, but their neurotoxic effects will not be considered here.

The classic symptoms of exposure to metallic mercury vapour are due to CNS involvement, while the kidney is the target for Hg(I) and Hg(II) salts. There is general agreement that the differences in toxicity between inorganic mercury compounds are largely determined by their physical properties and redox potentials. This section will exclusively be concerned with the behavioural toxicity of elemental mercury in occupational exposure settings.

Well-designed studies using matched controls and computerized psychometric tests, tremor tests, and other assessments such as mental arithmetic ability, perceptual speed, attention, choice reaction time, and finger tapping come to the same overall conclusion that exposure to mercury at the workplace produces both behavioural performance changes and altered mood states, and that these occur at low levels of exposure. Because of the cross-sectional character of all of the available studies there are inconsistencies, but these may be due to inadequate matching.

17.7.2.3 Manganese.   Manganese is an essential trace element which, in cases of excessive exposure, induces signs and symptoms of CNS involvement similar to those of Parkinson's disease. Most current knowledge of the neurotoxicity of manganese comes from neurological and behavioural studies in occupationally exposed workers. Contrary to lead and mercury, no biochemical markers are available to relate the target dose to neurotoxicity after long-term exposure. Instead, manganese concentration in dust is used to quantify workplace exposure.

Steel smelting workers with moderate dust exposure (190–1390 mg m^{-3} total dust) were compared with matched controls. No group differences were observed for most neurophysiological, psychiatric, and cognitive measures. However, the steel workers exhibited slower diadochokinesis, prolonged reaction times, and impaired tapping and digit span performance (short-term memory). In another study of workers from a dry alkaline battery factory exposed to manganese dioxide dust (948 mg m^{-3}), highly significant group differences were observed when compared with matched controls in sensorimotor performance, namely impaired eye–hand co-ordination, prolonged reaction times, and impaired steadiness (tremor). As a consequence the authors proposed a drastic revision of the currently accepted TLV for manganese, as well as regular surveillance of exposed workers, with simple sensorimotor tests for early detection of deficit.

17.8   FIELD STUDIES: ENVIRONMENTAL EXPOSURE

Behavioural research dealing with the neurotoxic effect of environmental pollutants has emphasized the developmental perspective. Infants and children have been the subjects of such studies and the focus of interest has been on inorganic lead as well as on PCBs.

17.8.1 Lead

Acute symptomatic lead poisoning is often associated with encephalopathy which, if the victim survives, may give rise to neurological and psychological sequelae. This is particularly true for children. Such clinical findings have led to the hypothesis that long-term low-level exposure during infancy and childhood may result in a subclinical neurobehavioural deficit even in asymptomatic children.

Many field studies in children have been reviewed by Pocock *et al.*, (1994) and the WHO. In these studies a variety of psychological tests have been used to assess the degree of CNS involvement, and both blood and tooth lead levels have served as internal markers of current or past environmental exposure. The dependent variables used in these studies included psychometric intelligence assessed in most studies by means of the Wechsler IQ Scales, although other tests such as McCarthy Scales of Children's Ability and British Ability Scales have also been used. Other functional domain tests have been added to the psychometric intelligence assessment such as perceptual motor functions, gross or fine motor co-ordination, sensorimotor speed and attention, as well as aspects of educational attainment.

Summarizing the outcome of many such cross-sectional approaches, which differed in terms of study design and confounder structure, the conclusion has been that doubling the blood lead concentration from 0.5 to 1.0 μmol l^{-1} is associated with a small but highly significant IQ deficit of around two points. The same holds true if doubling of tooth lead concentrations is considered. Results from the European Multicentre Study on Lead Neurotoxicity in Children (Winneke *et al.*, 1990) suggest that more elementary measures covering attention and perceptual motor functions may be more consistently impaired by lead than the complex IQ measure.

Prospective studies have also attempted to clarify the role of environmental lead exposure as a contributing factor in developmental neurotoxicity. These have also been reviewed by Pocock and the WHO. The strength of these studies lies in their prospective design, and the fact that initial agreement was reached on the main features of the core protocol covering independent, dependent, and main confounding variables. Neurobehavioural tests included the Bayley Scales of Infant Development until the age of 2 years, the McCarthy Scales of Children's's Ability beyond age 3, and the Wechsler IQ scales at or beyond 5 to 6 years of age.

Quantitative conclusions based on meta-analysis of these studies were similar to those drawn from the cross-sectional studies. A doubling of the blood lead level from 0.5 to 1.0 μmol l^{-1} was collectively associated with a statistically significant IQ drop of about two points. There was no evidence that pre- or neonatal lead exposure was associated with adversity beyond the age of 2 years. For both the cross-sectional and the prospective studies no effect threshold was identified and, as is true for observational studies in general, the real cause–effect contingencies must remain uncertain. Supporting evidence in this respect comes from behavioural toxicity studies in animals at comparable levels of internal exposure.

17.8.2 Polychlorinated Biphenyls

PCBs are mixtures of congeners which differ in the number and position of chlorine atoms on the two rings. Although no longer used or produced in most countries they are still of environmental concern in the context of waste disposal and because of their persistence in the environmental media. Lacking biodegradability they accumulate in the food chain and finally reach the human biosystem with elevated levels in different tissues, notably human milk.

In a comprehensive review covering experimental and epidemiological findings, the effects of PCBs on the nervous system have been emphasized. In epidemiological studies lower Bayley scores for psychomotor development in infants from North Carolina, and impaired visual recognition memory in 7-month-old infants from fish-eating families around Lake Michigan were found to be associated with PCB levels in cord serum rather than with postnatal PCB exposure *via* maternal milk. Whereas the memory deficit in the Michigan cohort persisted until at least age 4 years, this was not true for psychomotor deficit in the North Carolina cohort. Four- to seven-year-old children born to mothers who had experience of PCB exposure through contaminated rice oil only during pregnancy scored five points lower on psychometric IQ tests relative to carefully matched children born to unexposed mothers. From these observations the conclusion has been drawn that PCB-induced neurobehavioural toxicity may be long lasting and of prenatal rather than postnatal origin. This conclusion was supported in a cross fostering study in PCB-exposed rats (see Figure 17.3).

17.9 CONCLUSIONS

The administration of neurobehavioural tests covering cognitive and sensorimotor functions in experimental and epidemiological studies in children and adults has helped to document the risk of neurotoxicity associated with occupational or environmental exposure to low levels of solvents, neurotoxic metals, and PCBs. In many instances such observations have been used to establish or revise exposure limits, particularly in occupational settings.

It should be pointed out, however, that frequently such observations have not been sufficiently consistent to be fully acceptable within the scientific community and/or to regulatory bodies. This is particularly true, for example, in the case of lead in the environment. The critical neurobehavioural effects are small and embedded in a complex background of other confounding potential effects. In such cases, questions are still raised as to the true cause–effect contingencies. In these situations, behavioural studies in experimental animal models have helped substantially to corroborate field observations, to contribute to elaborating the underlying neurobiological mechanisms, and to identify critical periods of exposure.

17.10 BIBLIOGRAPHY

K. Anger, 'Human neurobehavioural toxicology testing', in 'Behavioural Measures of Neurotoxicity', eds. R.W. Russell, P.E. Flattau, and A.M. Pope. National Academy Press, Washington, DC, 1990, pp. 69–85.

M. Berlin, 'Mercury', in 'Handbook on the Toxicology of Metals', eds. L. Friberg, G.F. Nordberg, and V.B. Voik. Elsevier, Amsterdam, 1986, pp. 387–435.

R.B. Dick and B.L. Johnson. 'Human experimental studies', in 'Neurobehavioural Toxicology', ed. Z. Annau. Johns Hopkins University Press, Baltimore, 1986, pp. 348–390.

F. Gamberale, 'Application of psychometric techniques in the assessment of solvent toxicity', in 'Safety and Health Aspects of Organic Solvents', eds. V. Riihimäki and U. Ulfvarson. Alan R. Liss, New York, 1986, pp. 203–224.

V.G. Laties and R.W. Wood, 'Schedule controlled behaviour in behavioural toxicology', in 'Neurobehavioural Toxiciology', ed. Z. Annau. Johns Hopkins University Press, Baltimore, 1986, pp. 69–93.

D.S. Olton and R.J. Samuelson, 'Remembrance of places passed: Spatial memory in rats', *J. Exp. Psychol., Anim. Behav. Proc.*, 1976, **2**, 97–116.

S.J. Pocock, M. Smith, and P. Baghurst. Environmental lead and children's intelligence: A systematic review of the epidemiological literature, *Br. Med. J.*, 1994, **309**, 1189–1197.

V. Riihimäki and U. Ulfvarson, eds. 'Safety and Health Aspects of Organic Solvents'. Alan R. Liss, New York, 1986.

A. Sahgal, ed. 'Behavioural Neuroscience. A Practical Approach'. IRL Press,

Oxford: 1993, vol. 1. (Recommended chapters: A. Sahgal, Passive avoidance procedures, pp. 49–56; J.P. Aggleton, Behavioural tests for the recognition of non-spatial information by rats, pp. 81–93; J.N.P. Rawlins and R.M.J. Deacon, Further developments of maze procedures, pp. 95–106; C.A. Stewart and R.G.M. Morris, The watermaze, pp. 107–122).

B.T. Stollery, H.-A. Banks, D.E. Broadbent, and W.R. Lee, Cognitive functioning in lead workers, *Br. J. Ind. Med.*, 1989, **46**, 698–707.

E.L. Thorndike, 'Animal Intelligence'. Macmillan, New York, 1911.

H.A. Tilson and G.J. Harry, 'Behavioural principles for use in behavioural toxicology and pharmacology', in 'Nervous System Toxicology', ed. C.L. Mitchell. Raven Press, New York, 1982, pp. 1–27.

A. Wennberg, A. Iregren, G. Struwe, G. Cizinsky, M. Hagman and L. Johansson, Manganese exposure in steel smelters a health hazard to the nervous system, *Scand. J. Work Environ. Health*, 1991, **17**, 255–262.

WHO, 'Environmental Health Criteria No 165 – Inorganic Lead', International Programme on Chemical Safety, World Health Organization, Geneva, 1995.

G. Winneke, A. Brockhaus, U. Ewers, U. Kramer, and M. Neuf, Results from the European multicentre study on lead neurotoxicity in children: Implications for risk assessment, *Neurotoxicol. Teratol.*, 1990, **12**, 553–559.

CHAPTER 18

Ecotoxicity

MARTIN WILKINSON

18.1 INTRODUCTION

This overview introduces ecology to scientists with little or no previous knowledge of biology. Toxicology is most commonly concerned with effects of toxicants on humans. By contrast, ecotoxicology is concerned with effects on organisms other than man. This has three dimensions: toxicity to single species other than man, toxic effects on inter-relationships between species, and accumulation of toxicants by organisms and their movement between organisms and species. This requires a basic knowledge of ecology before the toxic effects can be fully understood. Following an introduction to ecology and to the unifying concept of a balanced ecosystem, this chapter examines in general terms the effects of man on ecosystems and the methods for monitoring ecological effects.

An understanding of ecotoxicology requires knowledge of how organisms interact in nature with each other (the biotic environment) and with the physical and chemical aspects of the environment (the abiotic environment). This is the science of ecology which can be viewed at several levels of organization, at each of which there can be toxic effects. Some examples of these levels in ascending order of complexity are given in Table 18.1.

The following account of ecology illustrates how these and other toxic effects can occur. A good starting point is an understanding of the sustainability of the ecosystem level. This requires considerable explanation which is now given in a very simple way assuming no previous knowledge of biology.

18.2 UNDERSTANDING HOW ECOSYSTEMS WORK

To start we are going to make a very simple assumption that there are two requirements that organisms have from the environment to sustain their life which take precedence over all other requirements: (i) a supply of carbon to form the organic molecules of which organisms are composed; (ii) a supply of energy to power the chemical reactions that keep the organisms alive. Carbon is freely available in the environment as carbon dioxide in the air and in various inorganic forms, including hydrocarbonate, dissolved in water. But organisms require

Table 18.1 *Levels of consideration in ecology*

Level of organization	Description of level	Examples of toxicant effects
1. Individual organism or species	Concerned with how physical and chemical environmental factors control which species can occur in which place	Alteration of the physical and chemical factors can affect the growth or survival of particular species
2. Population	A group of individuals of a single species living together and having interrelationships through gene exchange by sexual reproduction	Effects on population size; adaptation to toxicants by tolerant mutants spreading through population
3. Community	A collection of populations of different species living together in one place (habitat) giving species assemblages characteristic of particular conditions, *e.g.* oak woodlands	Changes in species composition owing to selectively different effects of toxicants on different species
4. Ecosystem	Organisms in a particular habitat considered together with their physical and chemical environment, and the processes linking the organisms and environment such as energy and nutrient flow and biogeochemical cycles. Ecosystems are characterized by a degree of sustainability	Interference with nutrient recycling; concentration and accumulation of toxic substances in food chains; alteration of productivity; sustainability can be impaired by these alterations

organic carbon and they can be divided into two major groups according to how they ensure their supplies of this, as shown in Table 18.2.

18.2.1 Carbon and Energy

In terms of number of species, autotrophs are very much in the minority, but they are of absolutely crucial importance because they make the organic matter that all organisms need. By far the biggest group of autotrophs, responsible for most of the fixation of inorganic carbon into organic form on the earth, are the plants which use the process of photosynthesis which can be summarized as follows:

$$6CO_2 + 6H_2O + \text{light energy} \rightarrow C_6H_{12}O_6 + 6O_2$$

This equation is an oversimplification of many reaction steps but illustrates the basic principle. The other fundamental process, respiration, is a series of breakdown reactions which, unlike photosynthesis, are undertaken by all organisms:

$$C_6H_{12}O_6 + 6O_2 \rightarrow 6CO_2 + 6H_2O + \text{chemical energy available for use in the cell}$$

Table 18.2 *Nutritional types of organism*

Type of organism	*Means of getting carbon*	*Means of getting energy*
Heterotrophic *e.g.* animals, fungi, some bacteria	**Ready-made organic carbon** By ingesting ready-made organic matter in the form of other living organisms or their waste products. Digestion to smaller molecules provides the building blocks for synthesis of other larger organic molecules using energy from respiration	**Chemical energy** By breaking down (catabolism) some of the larger organic molecules ingested in the diet in the process of respiration and applying the chemical energy released to synthesis (anabolism) of other chemicals needed by the organism
Autotrophic Mainly plants but also some bacteria	**Inorganic carbon** Carbon dioxide (on land) or hydrogen carbonate and other dissolved forms (in water) are reduced to organic carbon, primarily by photosynthesis in plants. Sugars resulting from photosynthesis can then provide an energy source in respiration or be used to synthesize other organic molecules	**Light energy** A physical form of energy, freely available in the environment, light, powers the anabolic reactions of photosynthesis in plants and some bacteria (but in a limited number of chemosynthetic bacteria chemical energy from inorganic reactions is used to reduce inorganic to organic carbon)

The living cell couples anabolic and catabolic reactions to transfer energy from one type to the other.

Only autotrophs make new organic matter, while all organisms consume it. Hence growth of new body matter of autotrophs is called primary production. Production of heterotrophs which simply recycle already existing organic matter is called secondary production. Therefore the production by the autotrophs must be sufficient to meet the needs of both autotrophs and heterotrophs for respiration. So in a balanced system there is a balance between production and respiration, and the photosynthesis by plants will be approximately balanced by total community respiration.

18.2.2 Inorganic Nutrients

Energy and carbon alone are not enough for life. About 20 different inorganic nutrient ions are needed because of their roles in biochemical reactions in living cells or because they are components of particular organic compounds, *e.g.* nitrogen in proteins. Plants absorb these from water and soil and they are passed to heterotrophs in the diet. Some nutrients, *e.g.* nitrogen and phosphorus, may often be in low concentrations in the environment compared with the amounts needed and so may limit plant growth and primary production. Other nutrients such as various metal ions may be even less abundant but are needed in much smaller amounts. Some such trace elements, *e.g.* copper, may be toxic when available in more than trace quantities but availability in soil or water may be regulated by natural chelators so reducing toxicity.

18.2.3 Food Chain

Organisms can be placed in a chain of dependence, known as a food chain, with several different trophic levels (levels at which organisms feed), with plants or primary producers absorbing light, inorganic carbon and nutrients, and passing nutrients and organic molecules with their chemical energy to the higher trophic levels of herbivores and carnivores (Figure 18.1). Each trophic level produces waste material (as excretory products and dead matter) and carbon dioxide from respiration. The waste products are broken down by decomposer organisms (bacteria and fungi) which release nutrients back to the environment where they are available for re-use. Therefore we can recognize that nutrients cycle between organisms and the environment. This is part of a more complex cyclic system, the biogeochemical cycle. For each element utilized by organisms there is such a cycle. The precise details differ between elements depending on the amount of the element available, the uses to which organisms put it, where they store it in their bodies, and the sinks for it in the environment. But all biogeochemical cycles incorporate the idea that, for any essential element at any one time, part

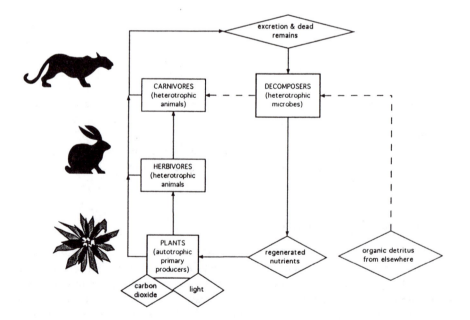

Figure 18.1 *Diagrammatic representation of a food chain. Organisms are contained in boxes and environmental requirements are shown in diamonds.*

of the total naturally occurring amount of the element is in the organisms and part is in different components of the natural environment. Individual atoms or molecules move between these compartments but the proportions in the different compartments remain roughly constant. These cycles must continue to function to ensure a supply of nutrients for organisms and to ensure continuing biological productivity.

Some organisms accumulate certain elements and compounds from the environment (bioaccumulation) causing them to have very high body loads relative to the outside concentrations (bioconcentration), *e.g.* metals in plant tissues. If the accumulated substance is conservative (not broken down by cellular processes) and stored, then a high dose will be given to the organisms that eat the bioaccumulator. Because of losses of organic matter due to respiration, each successive trophic level often has a lower biomass (mass of living material in a given area at one time) or productivity than the levels below it. The body concentration of conservative substances passed up the food chain can therefore increase up the chain (biomagnification), sometimes resulting in toxic doses to organisms higher up the chain.

Nutrients and carbon are recycled. The only requirement not recycled directly is light energy. Energy is lost to the environment by organisms as heat. Consequently, primary productivity is dependent on the continuous input of energy from the sun. It is also controlled by the availability of all the other requirements for plant growth, carbon dioxide, water, and nutrients. Since the availability of all these substances differs between habitats, different levels of primary production are characteristic of different places (Table 18.3).

The rate of secondary production depends on the availability of energy, carbon, and nutrients from the primary producers, so factors affecting plant growth usually affect total production of the whole system. An exception is some detritus-based systems such as estuaries. In estuaries the hydrographic conditions cause suspended particles from land drainage, the sea, and fresh-water to accumulate, giving turbid water which restricts light penetration for photosynthesis. The accumulated suspended matter includes much organic detritus which is used instead as a carbon, energy and nutrient source by estuarine heterotrophs. There is so much detritus that there is high secondary production despite restricted photosynthesis in this system. The primary production has been done in other habitats from which the detritus has been transferred.

A food web is a more realistic concept than a food chain. Figure 18.2 presents a very simple food web based on imaginary species (most natural ones would contain many more species). Even with such a simple one there can be a complex pattern of flow of energy, carbon, and nutrients, based on the feeding preferences of different species, as indicated by the lines on the diagram. For any particular habitat there is a degree of stability by which the same assemblages of species are present in a food web in successive years, with the same dominant and rare species, with the same flow pathways important and others less so. But what is it that determines which species will be present?

Table 18.3 *Generalized productivity of different habitat types*
(After Odum, 1985)

Habitat type	Gross productivity (g of dry matter m^{-2} per day) indicative of primary productivity
Deserts	> 0.5
Grasslands, deep lakes, mountains, forests, some agriculture	0.5–3.0
Moist forests and secondary communities, shallow lakes, moist grasslands, most agriculture	3–10
Some estuaries, springs, coral reefs, terrestrial communities on alluvial plains, intensive year-round agriculture	10–25
Continental shelf waters	0.5–3.0
Deep oceans	> 0.5

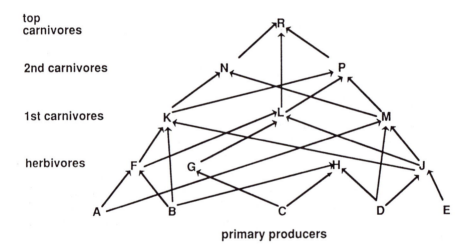

top carnivores

2nd carnivores

1st carnivores

herbivores

primary producers

Figure 18.2 *Theoretical food web for a group of imaginary species, indicated by letters. Lines joining the imaginary species indicate feeding relationships and hence pathways for the flow of energy, carbon, and nutrients. Note the complexity of the diagram since species exhibit feeding preferences rather than feeding on all the species on the trophic level below, and also because some species feed at more than one trophic level*

18.2.4 Environmental Gradient

Organisms do not occur together wholly by chance. A particular habitat has its own set of environmental conditions to which an organism must be tolerant if it is to exist. Different species have different tolerances to physical and chemical environmmental factors (abiotic factors), *e.g.* temperature, rainfall, soil nutrient status. The range of abiotic factors tolerated along a gradient of such factors (Figure 18.3) can be considered as the theoretical niche of the species. In practice, species usually occupy a narrower range of conditions than this, the realized niche. They do not occur at the extremities of the theoretical range because of interactions there with other organisms (biotic interactions). For example a species will be best adapted to the environment near to the middle of its tolerance range. Towards the extremities it might be under some stress. It will not compete there with other better-adapted species, which are towards the middle of their tolerance ranges.

18.2.5 The Ecosystem

We are now in a position to arrive at a concept of an ecosystem. It consists of all the organisms in a particular place or habitat, their inter-relationships with each other in terms of nutrients, carbon, and energy flows, and in terms of biotic determinants of community composition such as competition between species, the physical habitat, and the abiotic factors associated with it, which also play a role in determining community composition and primary, and therefore secondary, production. This may seem a woolly concept to a physical scientist who is used to a rigid quantitative approach. This is only a preliminary descriptive account. Ecosystems can be quantified, for example, in terms of the fluxes of carbon, energy, and nutrients, and the productivities of each trophic level. They can be quantitatively modelled using computers to enable predictions to be made about ecosystem performance.

For our purposes the important feature of ecosystems is their dynamic stability; that they are able to remain broadly constant over time in species composition

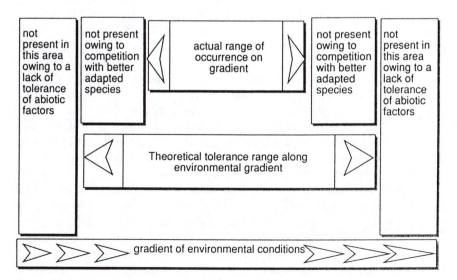

Figure 18.3 *Diagram to show the relationship of actual to theoretical tolerance ranges for a hypothetical species along a gradient of environmental conditions*

and abundance and in the magnitudes of processes, despite environmental variations. Although the climate fluctuates from year to year, the structure of the ecosystem tends to be stable within limits, and therefore it is sustainable. One example of dynamic stability is in population sizes. Man's population does not fluctuate wildly from year to year because the generation time is about 20 years and several generations are overlapping. In contrast, many insects reproduce every year and the life span is only 1 year or less. Here, there can be fluctuations of several orders of magnitude in population size over several years but they fluctuate around a mean value. This may result from density-dependent factors, environmental factors whose intensity or effect depend on the population density. For example, at high density food may be short, giving a population crash, whereas at low density the abundance of food may allow population size to increase, thus fluctuating over several years about a mean.

Ecosystem stability is not rigid. Some systems change naturally, hence the dynamic nature of the stability. On a short time scale this happens with winter and summer aspects of a community in a temperate climate. On a longer time scale there is ecological succession, where one community naturally replaces another on an area of land or water, usually as a result of the modification of the habitat conditions by the organisms that are replaced so that it is no longer suitable for their own survival. This particularly happens where an open area of land or water is available for colonization. An example is the formation and growth of maritime sand dune systems. Near the high tide mark on a beach is an inhospitable environment for plants, windswept with high water loss by evaporation and with sand abrasion, high sand surface temperatures in summer, and a low nutrient and highly saline soil, subject to erosion by waves and wind. Only a few species, the dune-building grasses, can tolerate this environment, forming an open community where, unusually, most ground area is not colonized. These grasses grow best through depositing sand which they stabilize, so building up dunes. The dune soil becomes less saline due to leaching by rain water, nutrients accumulate from the grass litter aided by nitrogen-fixing bacteria associated with their roots, and the growing dunes provide shelter. Going inland the habitat becomes progressively more normal, less inhospitable, and there is a progressive replacement of the dune-building grasses by a wider range of more normal, less tolerant plants. Ultimately a closed (complete ground coverage) climax community is achieved in equilibrium with the climate and with any local conditions such as soil type.

Mature, stable ecosystems are characterized by a preponderance of organisms referred to as K-strategists, species that succeed by being well-adapted to their environment. Earlier stages in a succession may have a greater proportion of r-strategists, organisms with wide environmental tolerance that do not survive so well in stable habitats in competition with more precisely adapted K-species. By contrast r-strategists are highly reproductive, flooding the environment with their propagules, ready to opportunistically colonize any habitat space that may become available. In stressful environments, either man-made stress or naturally harsh conditions, tolerance to abiotic factors becomes a greater determinant of community composition than biotic interactions, and r-strategists predominate.

The above description of the ecosystem concept has stressed the ability of such systems to remain stable within limits in various ways. Maintenance of this stability is the key to understanding effects on ecosystems due to pollutants.

18.3 MAN'S EFFECTS ON ECOSYSTEMS

Man affects the dynamic balance of ecosystems in two ways, by pollution and by disturbance. As this monograph is concerned with toxic effects, only pollutants will be considered here. Pollutants are hard to define, but will be regarded in this chapter as substances that can potentially have an impact on ecosystems, either because they are novel chemicals synthesized by man, which normal decomposer organisms are not accustomed to dealing with, or because they are discharged in unusually high amounts and/or to a system from which they did not come, *e.g.* human waste from food grown on land discharged in concentrated form through sewer outlets to rivers or the sea.

Ecosystems become unbalanced through pollutant effects, *i.e.* the stability is disturbed and the productivity and recycling impaired, meaning that they are no longer sustainable systems. This results from the selective action of toxicants, affecting different species in different ways, or to different extents, or at different concentrations. There may be lethal effects where species are killed but more commonly there are sublethal effects where species remain alive but with reduced growth or reduced reproductive ability or modified development, all leading to ecosystem alteration. A summary of ways in which toxic pollutants may affect organisms at the different levels of consideration in ecology is given as a flow diagram in Figure 18.4.

At an ecosystem level the above effects can give rise to various symptoms of stress in the system. However stress can be due not only to toxicants but also to non-toxic pollutants, to physical disturbance, and to natural stress in extreme habitats. Part of the art of measuring biological effects of pollution, which is summarized below, is in distinguishing man-made from natural stress effects. The symptoms of stress in ecosystems are given in Table 18.4.

As mentioned above, not all pollutants are directly toxic. Nonetheless some of the non-toxic ones are relevant to this account because they can have a secondary toxic effect. An example is enrichment of a water body with plant nutrients such as nitrogen and phosphorus (eutrophication) which can enter as pollutants from sewage, fertilizer run-off, or some industries. Assuming adequate supplies of carbon and light, plant growth will be limited by nutrients. Nutrient pollution then can have a fertilizing rather than a toxic effect. Considerable enrichment can give massive unchecked growth of plants which outstrips the ability of herbivores to graze on it. The decay of the excess plant biomass by bacterial activity then creates a demand for oxygen for bacterial respiration which may exceed its rate of supply from the overlying atmosphere. The resulting deoxygenation of the water can then have a lethal effect on animals, which require respiratory oxygen more critically than plants, which can produce their own oxygen through photosynthesis. Some of these effects on ecosystems can be used in biological measurements of pollution. The next section gives an overview of such techniques.

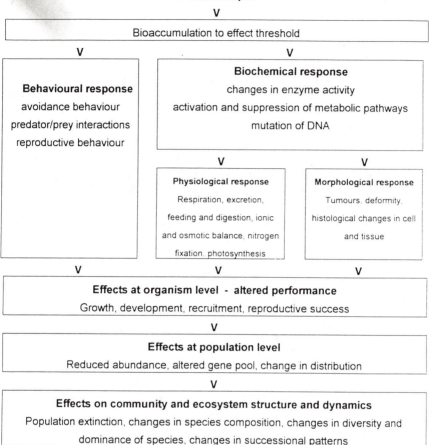

Pollutant input

V

Bioaccumulation to effect threshold

| V | V |

Behavioural response
avoidance behaviour
predator/prey interactions
reproductive behaviour

Biochemical response
changes in enzyme activity
activation and suppression of metabolic pathways
mutation of DNA

V | V

Physiological response
Respiration, excretion,
feeding and digestion, ionic
and osmotic balance, nitrogen
fixation. photosynthesis

Morphological response
Tumours. deformity.
histological changes in cell
and tissue

V | V | V

Effects at organism level - altered performance
Growth, development, recruitment, reproductive success

V

Effects at population level
Reduced abundance, altered gene pool, change in distribution

V

Effects on community and ecosystem structure and dynamics
Population extinction, changes in species composition, changes in diversity and
dominance of species, changes in successional patterns

V

Effects on ecosystem function
Reduced organic decomposition alterations in nutrient cycles, reduced primary
productivity

Figure 18.4 *Flow chart to show some of the ways in which toxic pollutants can affect natural
ecosystems at different levels*
(After Sheehan *et al.*, 1984)

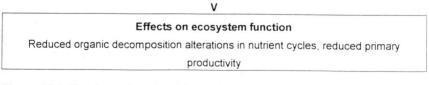

18.4   MEASUREMENT OF TOXIC EFFECTS ON ORGANISMS
AND ECOSYSTEMS

Measurement can be made by direct toxicity assessment or by ecological
monitoring, *i.e.* assessment of ecosystem effects.

Direct measurement or toxicity testing is a laboratory procedure carried out
with a single species using toxicants as single chemicals or as effluents before or

Table 18.4 *Trends expected in stressed ecosystems*
(After Odum, 1985)

Energetics
1. Community respiration increases
2. Production to respiration ratio becomes unbalanced
3. Primary production exported to other systems or remaining unused increases

Nutrient cycling
4. Nutrient turnover increases
5. Horizontal transport of nutrients (*i.e.* to other systems) increases
6. Vertical cycling (*i.e.* internal recycling) of nutrients decreases
7. Nutrient loss increases

Community structure
8. Size of organisms decreases
9. Life spans decrease
10. Species diversity decreases and dominance increases
11. Food chains become shorter

Ecosystem-level trends
12. Ecosystem becomes more open (*i.e.* more space available for colonization)
13. Successional trends reverse
14. Efficiency of resource use decreases

after mixing with the receiving environment. The organism is incubated under standard conditions for a fixed time in various dilutions, or with various doses of the toxicant, and with controls with no added toxicant. The concentration that brings about death of 50% of the individuals in the test population is the median lethal concentration (LC_{50}). Alternatively the single added dose that brings about 50% mortality is the median lethal dose (LD_{50}). Such lethal toxicity tests are popular because they are straightforward to carry out but they are an extreme measure.

Most toxic effects are sublethal and so sublethal tests should be used. This can be done by means of the median effective concentration (EC_{50}). This is the concentration of added toxicant which, in a given time under given conditions, brings about a 50% specified sublethal response, for example, a 50% reduction in growth rate relative to a control with no added toxicant. It could also be a 50% change in any sublethal measurement of a physiological process, such as a 50% reduction in photosynthetic or respiratory rate relative to a control, or a 50% change in a developmental process such as the formation of reproductive bodies.

A more stringent measure is the no observed effect concentration (NOEC). This is the highest concentration of added toxicant that has no measurable inhibitory sublethal effect on the test organism under the specified conditions in the prescribed time.

Results of toxicity tests can be attractive to regulators because they give a single, easily determined and repeatable, numerical measure. But they should not be extrapolated out of context. Problems exist in the selection of suitable test organisms and in the extrapolation of toxicity test results to field conditions.

Test organisms should be chosen to represent all of the main trophic levels, an autotrophic one such as a plant, a herbivore, and a carnivore. Fulfilling these criteria alone is not enough. The particular species chosen should be appropriate to the environment where the toxicant is to be discharged. There is a tendency to use a restricted range of species that can be found in culture collections, which might have evolved over long periods of repeated subculturing so as not to have the same environmental response as the original isolate. An example of inappropriate choice that has occurred is the use of the marine oyster embryo bioassay to test a susbstance to be discharged to a freshwater river, presumably because the oyster embryo test was a newly developed test that was popular at the time. There is a need for the development of a wider range of tests.

Apparently inappropriate tests might be suitable, used as a standard reference, to rank the general toxicity of many different chemicals. This may enable the choice of the least toxic alternative for any particular process. What cannot be done easily from laboratory tests is the prediction of effects on the structure or functioning of ecosystems. It is inherent in the nature of a toxicity test that it is done under constant laboratory conditions, which cannot mimic the complex and fluctuating field environment and the biotic interactions between species in the field, and therefore cannot elicit the same response from the organisms and their assemblages. One approach being taken to remedy this is the development of *in situ* tests which are carried out in the field. The organism is grown captive in a polluted location and some measure of its growth, physiology, biochemistry, or survival is compared with similarly treated captive organisms in a similar but less polluted control environment. These methods are in their infancy and do not always find favour because of the undefined nature of the conditions and the uncertainty that the control environment is similar in all features except the pollution.

The above is not to decry the use of toxicity tests but simply to counsel their wise use. For example it is only recently that the British water industry has started to build toxicity criteria into consents given to discharge liquid effluents into watercourses or coastal waters. Previously the consents contained only physical and chemical limits on effluent composition. The addition of toxicity criteria makes them more effective for complex effluents where there might be synergistic effects between components or where there might be so many components that they were not all regulated in the consent. It is the total toxicity of the effluent that is being assessed rather than its composition of specific chemicals.

Enforcement of toxicity criteria in consents to discharge could be a problem. While toxicity tests are attractive to some because of their ease and simplicity, routine application of tests on a wide range of organisms with a large number of effluents could be very costly, especially if vertebrates are used, *e.g.* fish, which require complex facilities and Home Office approval. An alternative, quick screening technique has been devised based on bacterial luminescence, of which Microtox is one proprietary test. This is based on light emission by a culture of luminescent bacteria following activation from dormancy. This light emission is reduced when the bacteria are in toxic solutions, relative to a control in clean media. Hence an EC_{50} can be calculated in terms of a 50% reduction in luminescence relative to the control. This might be thought to be a particularly bad example of an inappropriate test organism but it is only being used as a first screening. If serious toxicity is shown in the relatively quick and cheap Microtox test then the more relevant but time consuming and expensive tests with the full range of organisms can be carried out.

Ecological monitoring is a broader assessment of the ecological effects of toxicants than is given by toxicity testing. It is defined as the assessment of effects of toxicants and pollutants in an ecological context, either by means of their accumulation in organisms other than man, or by looking for abnormal ecological effects at the level of species, community, or ecosystem. It performs a different role from that of chemical analysis of toxicants in the environment. Chemical analysis usually relies on occasional instantaneous sampling. It does not necessarily indicate average, maximum, or minimum environmental values of the toxicant. Ecological monitoring avoids the extremely frequent chemical sampling necessary to get over this problem. Indigenous organisms integrate concentrations of toxicant over time. Furthermore they show what chemical sampling cannot do (determine the effects of the toxicants on the natural communities),a reason for carrying out such monitoring. Ecological methods do not give numerical estimates of toxicant concentrations, so both chemical and ecological approaches should be taken.

Ecological monitoring can use naturally occurring organisms in the field or organisms transplanted to the field for the purpose, and may be supported by laboaratory tests. Table 18.5 presents an illustrative selection of approaches to ecological monitoring, with a bias towards aquatic assessment.

18.5 CONCLUSIONS

Toxic pollutants can disturb the sustainability of natural ecosystems by a variety of effects on species, populations, communities, and ecosystem processes, although such systems characterized by dynamic stability have some capacity to absorb pollutants. Toxicity testing has limitations in predicting such effects and chemical measurement of environmental toxicants should be accompanied by ecological monitoring. A difficulty requiring specialist knowledge is the distinction of ecological effects due to pollution, disturbance, and naturally difficult environmental conditions.

Table 18.5 *An overview of selected measures used in ecological monitoring*

1. *Assessments carried out in the field*

Using organisms occurring naturally in the environment	Pollutant accumulation by organisms (bioaccumulation monitoring)	Some organisms accumulate metals, radionuclides and some hydrocarbons to high levels in their tissues in proportion to the external concentration. Gives higher, more detectable concentrations. Integrates concentration over time. May indicate biologically active fractions of the substance. Algae may indicate dissolved fraction while animals feeding on suspended matter (*e.g.* mussels) may indicate particulate fraction
	Assessments using single species	Presence or absence of indicator organisms. There are few genuine indicators solely by presence so must be used with care
		Biochemical measurements on single species: measurement of activity or amount of substances induced by presence of pollutants, *e.g.* enzymes or metal-binding proteins
		Pathology: presence of tumours induced by pollutants
	Assessments using communities and populations	Age structure: in a species that can be aged and which recruits annually, abnormal age structure may indicate a failure to recruit in one year due to pollution or to natural climatic factors
		Life-forms and successions: successions regressed to earlier stages with abnormal abundance of opportunists may indicate stress
		Numerical structure:
		(i) Species richness: fewer species may occur under stress.
		(ii) Diversity: there are many numerical indices, which are mathematical formulations of species number, numbers of individuals, and the distribution of individual numbers between species. They are used as general assessments of community structure in ecology but variations from expected values can indicate toxicant-induced stress. Also specially developed indices are available such as the Trent Biotic Index which indicates degree of sewage stress on animal communities in rivers based on numbers of taxa and presence of key species or groups

Table 18.5 *(contd)*

Using organisms planted out at test site	*In situ* toxicity assessment using measurements of growth of organisms at a test site compared with a control site
	Colonization of artificial substrata: provides a uniform substratum that can be compared at different sites using numerical indices (see above) of the communities of small organisms that develop
	Colonization of cleared natural substrata, again using numerical indices of community structure, may also show whether an alternative community can develop under pollutant influence when the established one is dislodged
	Bioaccumulation monitoring using monitors artificially placed at a variety of test sites to enable comparison
2. Tests carried out in the laboratory	
Toxicity testing	LC_{50}, LD_{50}, EC_{50}, and NOEC—limitations as described earlier (see text)
Growth potential	Testing of survival or growth rate of organisms in laboratory culture under standard conditions in waters from test sites in comparison with water from clean control sites to see the extent to which test sites might support growth of certain sites
Biostimulation	Measurement of the growth of algae in water spiked with various concentrations of different added nutrients to determine its potential for eutrophication

18.6 BIBLIOGRAPHY

J.M. Anderson, 'Ecology for the Environmental Sciences: Biosphere, Ecosystems and Man'. Edward Arnold, London, 1981.

A. Beeby, 'Applying Ecology'. Chapman and Hall, London, 1993.

F. Moriarty, 'Ecotoxicology'. Academic Press, London, 2nd edn., 1988.

E.P. Odum, 'Trends in stressed ecosystems', *Bioscience*, 1985, **35**, 419–422.

P.J. Sheehan, D.R. Miller, G.C. Butler, and P. Bourdeau, eds. 'Effects of Pollutants at the Ecosystem Level'. SCOPE, Wiley, Chichester, 1984.

WHO, 'Environmental Toxicology and Ecotoxicology: Proceedings of the Third International Course', Environmental Health Series No 10, World Health Organization, Copenhagen, 1986.

CHAPTER 19

Radionuclides

MILTON PARK

19.1 RADIOACTIVITY

The phenomenon of the natural radioactivity of some chemical elements was first appreciated by Becquerel in 1896: he observed that photographic emulsions wrapped in black paper and placed near a uranium compound, potassium uranyl sulfate, were blackened. This effect was subsequently attributed to the emission of a radiation by the uranium with properties not dissimilar to those of the already known X-rays, in that it was capable of ionizing air, and the activity of a uranium compound could be measured by the rate at which a known quantity could bring about the discharge of an electroscope. The emission of these rays was a fundamental property of the uranium atom, the activity being independent of the nature of the compound, of its valence state, of the temperature, or of the previous history of the material. The spontaneous emission of radiation of this type is now known as radioactivity.

This ionizing radiation differs from non-ionizing radiation, such as light or radio waves, in its possession of sufficient energy to remove electrons from the atoms of matter through which it passes, and therein lies its particular hazard. This is in contrast to non-ionizing radiation, which does not normally possess this property.

Following Becquerel's discovery, the work of Rutherford and Soddy, and of Pierre and Marie Curie, established that the nuclei of some natural elements were not completely stable. These unstable elements were found to emit radiation of three main types, two having the properties of charged particles (alpha and beta radiation), and the third having the characteristics of high-energy electromagnetic radiation.

19.2 TYPES OF IONIZING RADIATION

The following types of ionizing radiation are associated with radioactivity.

19.2.1 Alpha Radiation (α)

Alpha radiation has been shown to be composed of helium nuclei, consisting of two protons and two neutrons bound together very tightly to give a very stable unit. Consequently each particle possesses a positive charge of 2 units, and a mass of 4 mass units.

These particles possess single kinetic energies of the order of several mega electronvolts (MeV) characteristic of the radionuclide emitting them. One electronvolt (eV) is defined as the energy gained by an electron passing through an electric potential of 1 volt. Because of their comparatively high charge, alpha particles interact intensely with matter, and in consequence impart energy to the medium along their path to a much greater extent than beta particles.

19.2.2 Beta Radiation (β)

This comprises high-speed electrons of kinetic energy up to more than 3 MeV originating in the nucleus. They are identical to atomic electrons in mass (1/1840 unit) and in charge (1 unit of negative charge). Although in normal use the term beta particles or radiation refers to these high-speed negative electrons, a further type of beta particle is also known. This has the same mass as an electron, but is positively charged, and is known as positron radiation. The particle is indicated by the symbol β^+ to distinguish it from the more common β^-.

An emitted positron ultimately combines with an electron, its anti-particle, resulting in the annihilation of both, and the conversion of their masses into annihilation radiation, which appears as two gamma-ray photons of energy 0.51 MeV. Positron emitters can be detected not only through their positron emission but also through this characteristic emission of these 0.51 MeV gamma photons. Sodium-22 (^{22}Na) and fluorine-18 (^{18}F) are examples of positron emitters and have found wide use in tracer studies.

The energies of particles from a single radionuclide exhibit a continuous distribution of energies from zero to a maximum value (E_{max}). Only a very small fraction of the emitted beta particles have energies close to the maximum, most having much lower energies: a representation of the distribution of the energies of beta particles from tritium (E_{max}= 0.018 MeV) is shown in Figure 19.1. Although the shapes of all beta spectra are broadly similar, their precise shapes and E_{max} values are characteristic of each radionuclide. The E_{max} value is the characteristic usually given to describe beta energies. However, in dosimetry a

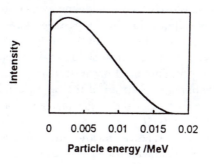

Figure 19.1 *The energy spectrum for tritium*

more useful quantity is the average energy. This quantity is a function of E_{max} and of the atomic number of the element, but it approximates to $E_{max}/3$.

19.2.3 Gamma Radiation (γ)

Gamma radiation is emitted only in conjunction with other types of decay, and belongs to the class known as electromagnetic radiation (like radio waves and visible light, but of very much shorter wavelength and higher energy). It is emitted when the nucleus produced following radioactive decay is in an excited state, and then returns to the ground state by emitting this radiation to carry away excess energy. Gamma energies range from a few keV to several MeV, and although one radionuclide may emit gamma rays of several different energies, their energies and relative intensities are specific to that nuclide and can be used to identify it.

19.2.4 X-rays

Similarly, X-rays are also a form of electromagnetic radiation, but differ from gamma radiation in that they result from extra-nuclear loss of energy of charged particles, for example electrons, but having shorter wavelengths than ultraviolet radiation. They may be emitted when an orbital electron of an atom jumps to another orbit of lower energy. The difference in energy is radiated as electromagnetic radiation. If the energy is high enough for the radiation to cause ionization, the emission is called an X-ray. Since the energy levels of X-rays are determined by differences between the energy levels of orbits, they have fixed values for any particular transition.

These vacancies in electron orbits can arise from the phenomenon known as electron capture, in which the nucleus captures one of the innermost K shell electrons, resulting in the conversion of a nuclear proton into a neutron. In filling the resulting vacancy in the K shell a rearrangement takes place with other electrons dropping to lower energy orbitals and emitting their excess energy as X-rays. A complex series of X-rays of different energies results, but one or two energies usually predominate.

A further source of X-rays is the phenomenon known as bremsstrahlung ('braking radiation'). This occurs when high-energy electrons are slowed down in an absorber: part of the energy lost by the particles is radiated as X-rays with a broad spectrum of energies. This is always present when beta particles are absorbed and can cause problems in shielding against beta radiation. Note that an X-ray tube is mainly an emitter of bremsstrahlung, its spectrum comprising a combination of a continuous spectrum of the bremsstrahlung produced by the braking effect of the electrons in the target material, for example copper, and X-rays produced by the orbital electrons of the target material dropping into vacant orbitals at lower energies, as mentioned above. These latter transitions occur at specific energies characteristic of the target element, and are known as the characteristic X-rays of that element.

19.2.5 Neutron Radiation

The final particle that will be mentioned here, although briefly, is the neutron. This is a very common particle, being a basic constituent of the nucleus and having a mass almost identical to the proton but carrying no charge. There are no significant naturally occurring neutron emitters, but radionuclides that emit neutrons can be produced artificially, and the neutron is of great importance both in nuclear fission reactors and in the production of radionuclides not available naturally.

19.3 RADIONUCLIDES

Each element can exist in the form of several nuclides, the same nuclide containing a given number of protons and neutrons in the nucleus, *i.e.* all atoms possess the same atomic number and mass number. Atoms possessing the same number of protons in the nucleus (which are therefore chemically identical and of the same element) but differing numbers of neutrons are called isotopes. Thus ^{12}C is a nuclide containing a nucleus composed of six protons and six neutrons. Hydrogen has three isotopes, from ^{1}H to ^{3}H.

Most naturally occurring nuclides are stable, but some possess unstable nuclei and transform spontaneously into the nuclide of another element, emitting radiation in this process. These unstable nuclides subject to radioactive decay are described as radionuclides. Unstable nuclides are more common in the heavier elements found in nature: all nuclides with an atomic number greater than 83 (bismuth) are radioactive. Of these, only uranium (^{235}U and ^{238}U) and thorium (^{232}Th) exhibit half-lives long enough to enable them to exist naturally. They decay by different sequential reactions, with emission of alpha and beta radiation, until the nuclei have broken down to stable nuclides of lead. Several lighter elements also have naturally occurring radionuclides, in particular ^{14}C and ^{40}K (Table 19.1).

Table 19.1 *Some examples of half-lives of radionuclides*

Radionuclide	^{14}C	^{60}Co	^{125}I	^{40}K	^{3}H	^{212}Po	^{90}Sr
Half-life	5730 y	5.27 y	60.1 d	1.28×10^9 y	12.3 y	2.98×10^{-7} s	29.1 y

^{40}K is the main source of radioactivity within the body; this radionuclide constitutes 0.0117% of the total potassium in the body. It is not being replenished, and presumably is the residue of material formed at the time the earth was created which still remains because of its very slow rate of decay. A different situation is found with ^{14}C. This decays at a much faster rate and is being continuously produced in the upper atmosphere by the interaction of cosmic-ray neutrons with nitrogen in the air: the nitrogen nucleus absorbs a neutron and releases a proton, resulting in its conversion into ^{14}C. This constant level of ^{14}C in the environment has been made use of in the technique known as radiocarbon dating.

Over the last 50 years or so, a number of radionuclides of stable elements have been produced by artificial means, and there is today a considerable industry for their production for a variety of scientific, medical and industrial purposes. Thus radioactive ^{60}Co can be made by irradiating non-radioactive ^{59}Co with neutrons in a nuclear reactor. The nucleus may capture a neutron with the emission of a gamma photon, referred to as a neutron, gamma (n, γ) reaction, resulting in its conversion into ^{60}Co. ^{60}Co is unstable and decays to the stable ^{60}Ni with the emission of a beta particle and gamma radiation:

$$^{59}_{27}\text{Co} + \text{n} \rightarrow {}^{60}_{27}\text{Co} \rightarrow {}^{60}_{28}\text{Ni} + \beta^-$$

19.4   THE UNIT OF RADIOACTIVITY

Radioactive decay is a random process, so it is impossible to predict when a particular nucleus will decay. However, since large numbers are involved, it is possible to forecast when a·proportion of the nuclei will have decayed. As a consequence of this random behaviour, radioactive decay is found to follow first-order kinetics, *i.e.* the rate of decay of a particular radionuclide is proportional to the number of nuclei of that nuclide present, or

$$-\frac{\mathrm{d}N}{\mathrm{d}t} = \lambda N$$

where N is the number of nuclei of the radionuclide present, and λ is the rate constant for this process, the radioactive decay constant. This equation also indicates that measurement of the rate of disintegration can be used to determine the amount of radioactive material present.

Integration of this equation between an initial time 0 and a time t, with N_0 the number of nuclei at time 0, gives,

$$N = N_0 \exp(-\lambda t)$$

The decay constant, λ, is characteristic of a particular radionuclide, but a quantity more commonly used for this purpose is the half-life, $t_{1/2}$ or $t_{0.5}$, of a radioactive species. This is defined as the time it takes for one-half of the nuclei in a sample to decay. In consequence of the first-order kinetics of the decay process, this, like the decay constant, is also a constant characteristic of a particular radionuclide. Its relationship to the rate constant can be shown by substituting the appropriate values of N ($N_0/2$) and t ($t_{1/2}$) in the above equation, to produce the equation:

$$t_{1/2} \times \lambda = \ln 2$$

Some examples of radioactive half-lives with values ranging from less than a microsecond to millions of years are given in Table 19.1.

Since the disintegration rate, or activity, A, is proportional to the number of unstable nuclei, we have the corresponding relationship between activity and time,

$$A = A_0 \exp(-\lambda t)$$

where A_o is the initial activity at time zero, or

$$A = A_0 \exp(-0.693t/t_{1/2})$$

in terms of ln 2 and $t_{1/2}$. Hence, if the half-life of the radionuclide is known, the activity of a sample can be calculated at any time (Table 19.2).

Table 19.2 *The decay of a sample of ^{32}P ($t_{1/2}$ = 14.29 days) with an initial activity of 100 disintegrations s^{-1}*

Time (days)	*Number of half-lives*	*Activity* (s^{-1})
0.0	0	100.0
14.3	1	50.0
28.6	2	25.0
42.9	3	12.5
57.2	4	6.3
71.5	5	3.1

The current SI unit of radioactive activity or quantity of radioactive material is the becquerel, 1 becquerel (Bq) corresponding to an activity of 1 nuclear disintegration per second (1 dps), which usually involves the emission of one or more charged particles (alpha or beta) and possibly X-rays or gamma radiation. An older unit which was also used is the curie (Ci), originally related to the activity of 1 g of radium. It is equivalent to 3.7×10^{10} Bq, or 37 GBq.

19.5 INTERACTION OF RADIATION WITH MATTER

Radiation can be divided into two classes: that comprising charged particles such as alpha and beta particles (directly ionizing radiation) and that comprising electromagnetic radiation and uncharged particles such as neutrons (indirectly ionizing radiation). They each interact differently with the medium through which they pass.

19.5.1 Directly Ionizing Radiation

Charged particles (*e.g.* alpha and beta) lose energy when passing through a medium mainly by interacting with the electrons of the medium. These are either excited to higher energy levels or are ionized from the parent atom. A further possibility is the emission of bremsstrahlung mentioned earlier. The charged particles emitted from radionuclides have only a very limited penetration range in the body, at most a few mm.

Alpha particles move at significantly slower speeds than beta particles; they are also much heavier (over 7300 times the mass of an electron) and possess double the electrical charge. Hence they impart energy to the medium at a markedly greater rate than beta particles, producing a much denser track of ionization along their path than an electron of equivalent energy. Consequently they have a much shorter range (in water the range of a 1 MeV beta particle is about 4.3 mm, whereas that of an alpha particle of the same energy is only about 7 μm).

The rate at which charged particles impart energy to a medium is known as linear energy transfer (LET). The LET values of particles (units typically keV μm^{-1}) are functions of their velocities and energies, increasing at lower energies and velocities. The LET of typical alpha particles is much greater than that of beta particles. A consequence of this is that because of their very limited range the hazard from alpha emitters outside the body (external exposure) is very small; however, the hazard is particularly great if they are inhaled or ingested, thereby placing them in close proximity to cell surfaces (internal exposure).

19.5.2 Indirectly Ionizing Radiation

X- and gamma rays, being uncharged, do not lose their energy in the same way. They can be considered as continuing until they collide with a nucleus or electron of the medium through which they are passing. All or part of the energy of the photon is then transferred to the particle. Collision with an electron can result in its being ejected from the parent atom with either complete absorption of the photon, or with absorption of only part of the photon energy and scattering of the beam with reduced energy. A further possibility is that, particularly close to a nucleus, an energetic gamma photon may be converted into a positron–electron pair.

Effectively, gamma and X-rays transfer all or part of their radiation to charged particles which then interact as discussed for alpha and beta particles above. However, they are much less attenuated as they pass through than are alpha or beta particles, and in consequence a proportion of their radiation may not be absorbed and pass right through a medium through which it is passing. A further consequence of this is that they can bring about ionizing events much deeper within a medium than can alpha or beta particles.

The deposition of energy from X- and gamma rays occurs along the tracks of the secondary electrons resulting from the absorption process. However, the degree of ionization is much less than occurs with alpha particles.

19.6   BIOLOGICAL EFFECTS OF IONIZING RADIATION

The particular hazards associated with radionuclides arise from their property of emitting radiation of sufficient energy to bring about the ionization of atoms in the medium through which they are passing. The first stage involves the generation of

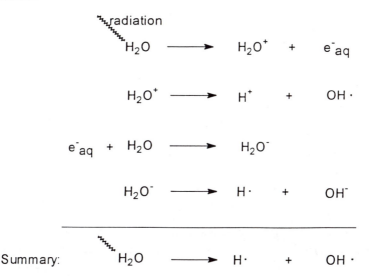

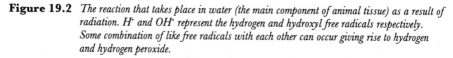

Figure 19.2 *The reaction that takes place in water (the main component of animal tissue) as a result of radiation. H· and OH· represent the hydrogen and hydroxyl free radicals respectively. Some combination of like free radicals with each other can occur giving rise to hydrogen and hydrogen peroxide.*

free radicals—highly reactive species bearing an odd electron that can react with and covalently modify other molecules, as illustrated in Figure 19.2.

In the next stage, or chemical stage, lasting only seconds, these free radicals can attack components of the cell, attaching themselves to molecules or causing breaks in long chain molecules. Of particular importance are their effects on the chromosomes and DNA of the nucleus, and on the permeability of membranes in general. At the low-dose levels associated with radiological protection, the most radiosensitive structure in the cell is the DNA of the nucleus.

In the final, or biological, stage, these chemical reactions may result, in simplified terms, in three outcomes as follows.

19.6.1 Death of the Cell

If sufficient cells are killed in the general irradiation of a person or animal, particularly the precursor (stem) cells providing a supply of mature functional cells in the blood, and gastrointestinal tract lining, skin, or gonads, the condition known as radiation sickness appears after a few hours or days. If the dose is sufficiently high, it will result in death.

19.6.2 Cell Survival but Permanent Molecular Modification

The most likely result in this case is modification of the cell DNA. This modification may be harmless, or it may give rise at a later stage in daughter cells to a malignant transformation, resulting in development of a cancer. If the damage occurs in a cell whose function is to transmit genetic information, such as cells in the reproductive organs, this may result in a hereditary effect, or defect, being passed on to future generations of the organism.

19.6.3 Repair

Living organisms have always been exposed to ionizing radiation from the natural environment, and there are repair mechanisms present in the cell to counter this type of damage. Particularly at low radiation doses, it is most probably repaired with no deleterious effects on the organism. However, the repair may be imperfect, resulting in a damaged but viable cell, as in section 19.6.2.

These repair mechanisms appear to be highly efficient. It has been estimated that many million million ion pairs are created in the total mass of the DNA of an individual in any year by the exposure of the body to natural background radiation alone. Despite this, the incidence of deaths from cancer is no more that 1 in 4, and only a small proportion of these cancer deaths is attributable to radiation. Thus the probability of one of these ionizations giving rise to a cancer is very small indeed.

19.7   UNITS OF RADIATION DOSE AND EQUIVALENT DOSE

Ionizing radiation cannot be detected by the human senses, but a number of instrumental methods are available for this purpose. These include photographic film, which is blackened by ionizing radiation, thermoluminescent material, Geiger tubes, and scintillation counters. Measurements made by such methods can be interpreted in terms of the radiation dose absorbed by the body, or a particular part of the body.

19.7.1 Radiation Absorbed Dose

The original unit of exposure was the röntgen (R), which was defined in terms of the radiation producing a certain amount of electrical charge in air. This has been replaced by the concept of radiation absorbed dose, a measure of the energy deposition in any medium by ionizing radiation. The SI unit of this is the gray (Gy), with 1 Gy corresponding to an energy deposition of $1\,J\,kg^{-1}$. An older unit of absorbed dose, the rad, is also found. This is approximately the absorbed dose in tissue exposed to 1 röntgen, and is equivalent to $0.01\,J\,kg^{-1}$. The unit, the centigray (cGy), which is equal to 1 rad, is sometimes used.

19.7.2 Equivalent Dose

It was found that the same amount of absorbed dose of different types and energies of radiation could produce very different amounts of biological damage, *i.e.* this was a function of both the absorbed dose and the type of radiation. Thus 1 Gy of alpha radiation was found to cause more tissue damage than 1 Gy of beta radiation, as a result of the much higher LET of alpha radiation. Consequently, when using different types of radiation, the equivalent dose is obtained by multiplying the absorbed dose by an empirical factor, the radiation weighting factor, w_R, the value of which is dependent on the type and energy of the radiation incident on the body, or in the case of sources within the body, emitted by the source. The unit of equivalent dose is the sievert (Sv), and this is related to the gray by the relationship,

$$\text{sievert} = \text{gray} \times \text{radiation weighting factor } (w_R).$$

The values of w_R are broadly related to the linear energy transfer (LET) of the different radiations. The values of the radiation weighting factors for some types of radiation are given in Table 19.3. All radiations of low LET have been given a value for w_R of unity, with that for other radiations being based on the observed values of relative biological effectiveness. This is defined as the inverse ratio of the absorbed doses producing the same degree of a defined biological effect.

Table 19.3 *Radiation weighting factors for some types of ionizing radiation.*

Type and energy range	Radiation weighting factor, w_R
Gamma radiation	1
X-radiation	1
Beta-particles	1
Alpha-particles	20
Neutrons, energy <10 keV	5
Neutrons, energy 10 keV to 100 keV	10
Neutrons, energy 100 keV to 2 MeV	10

Where a tissue, T, is irradiated with different types of radiation with different radiation weighting factors, the equivalent dose in that tissue (H_T) is obtained by summing the products of the absorbed dose in that tissue ($D_{T,R}$) and the radiation weighting factor for each type of radiation:

$$H_T = \sum_R w_R \times D_{T,R}$$

The equivalent dose thus provides an index of harm to a particular tissue from various radiations: *e.g.* 1 Sv of alpha radiation to the lung is deemed to create the same risk of inducing a fatal lung cancer as 1 Sv of gamma radiation, although the absorbed dose (Gy) is much greater in the latter case.

19.7.3 Effective Dose

A further complication is that different tissues exhibit different risks of the development of a fatal cancer: for a given dose of a particular type of radiation to the lungs and to the skin, the risk to the lungs is much greater than that to the skin.

This becomes of particular significance when irradiation is non-uniform, such as might be produced by a source of radiation within the body: only certain tissues near the source might be subjected to significant radiation dose. This is taken into account by summing the equivalent doses to each of the tissues of the body multiplied by a weighting factor related to the risk associated with that organ:

$$E = \sum_T w_T \times H_T$$

where E is the effective dose, H_T is the equivalent dose in a tissue or organ T, and w_T is the tissue weighting factor for tissue T. The tissue weighting factors are shown in Table 19.4. The sum of the tissue weighting factors has been normalized to unity. This allows a variety of non-uniform distributions of dose in the body to be expressed as a single number, broadly representing the risk to health from any of the different distributions of equivalent dose or from a similar dose received uniformly throughout the whole body. Table 19.5 summarizes the different dose quantities discussed above.

Table 19.4 *Tissue weighting factors for the calculation of the effective dose*

Tissue or organ	Tissue weighting factor, w_T
Gonads	0.20
Bone marrow (red)	0.12
Colon	0.12
Lung	0.12
Stomach	0.12
Bladder	0.05
Breast	0.05
Liver	0.05
Oesophagus	0.05
Thyroid	0.05
Skin	0.01
Bone surface	0.01
Remainder	0.05
Whole Body Total	1.00

19.8 EFFECTS OF RADIATION IN MAN

For some considerable time now it has been recognized that short-term harmful effects could be produced by over-exposure to ionizing radiation. It is only in the last 30 or 40 years that there has been a realization of the long term adverse effects of much lower doses of radiation, effects that are not related directly to dose.

The sources of information about the effects of irradiation on man are very meagre and unsatisfactory: the information is not obtained under carefully controlled conditions in most cases, rendering much of the data uncertain. These

Table 19.5 *Summary of dose quantities*

Dose quantity	Explanation and units
Absorbed Dose	Energy imparted by radiation to unit mass of tissue (Gy or J kg^{-1})
Equivalent Dose	Absorbed dose weighted for harmfulness of different radiations (radiation weighting factors) (Sv)
Effective Dose	Equivalent dose weighted for susceptibility to harm of different tissues (tissue weighting factors) (Sv)

sources include data from the atomic bomb victims of Hiroshima and Nagasaki, from victims of fall-out from nuclear tests, and from radiation accident and therapy cases.

For radiological protection, two types of health effects are often defined: deterministic and stochastic.

19.8.1 Deterministic Effects

With deterministic effects, the severity of the effect is dependent upon the dose, *e.g.* the acute radiation syndromes associated with substantial whole body irradiation. In these cases some threshold dose has to be exceeded before the effect becomes apparent. Examples of these effects include radiation sickness, cataracts, and damage to the skin.

Acute radiation syndromes following whole body exposure occur after exposure to doses of radiation well above those used in the setting of dose limits. Irradiation affects the precursor (stem) cell pools, particularly those tissues with a high turnover rate, such as bone marrow, the gut lining, skin, and germinal epithelium. Exposure to a few grays will result in a sudden loss of cell replacement capacity. From the meagre data available, no individuals would be expected to die at doses below 1 Gy, and the LD_{50} in 60 days ($LD_{50/60}$) for an acute exposure has been estimated to be between 3 and 5 Gy (Table 19.6).

19.8.2 Stochastic Effects

With stochastic effects, in contrast, the probability of the effect is dependent on the dose, and there is assumed to be no threshold. Examples of these are fatal

Table 19.6 *Range of doses associated with specific radiation-induced syndromes and death in human beings exposed to acute, low-LET, uniform, whole body radiation*

Whole body absorbed dose (Gy)	Principal effect contributing to death	Time of death after exposure (days)
3–5	Damage to bone marrow ($LD_{50/60}$)	30–60
5–15	Damage to the gastrointestinal tract and lungs	10–20
>15	Damage to nervous system	1–5

cancers and serious hereditary disease. It would appear that no stochastic effects other than cancer (and benign tumours in some organs) are induced by radiation in an exposed individual.

In contrast to the relatively high doses required before deterministic effects are noted, stochastic effects occur at much lower values of equivalent dose. Based on an analysis of the type of data already mentioned, the International Commission on Radiological Protection (ICRP) have estimated that for adult workers, assuming uniform irradiation, the probability of dying from a radiation-induced cancer is 4×10^{-2} Sv^{-1}. Thus for a person whose working life extended over 50 years and was subjected to an annual dose of 10 mSv in the course of his work, *i.e.* a cumulative dose of 0.5 Sv, the probability of death from cancer attributable to radiation exposure is $0.5 \times 4 \times 10^{-2}$, *i.e.* 0.02 or 2%. This corresponds to an annual risk of 1/50th of that, *i.e.* 0.04% or 1 in 2500. At this dose rate no deterministic effects would be apparent, although the risk of a fatal cancer would be significant. Obviously, if the annual dose were only 1 mSv year^{-1} the risk would be correspondingly reduced. To place these risk factors in perspective, the annual risk of death from smoking is 1 in 200, from natural causes in a 40 year old, 1 in 700, from accidents on the road, 1 in 10 000, and from accidents at work (all employment), 1 in 50 000.

The current exposure of the general public in the UK to different sources of radiation is given in Table 19.7. Of the total annual effective dose of 2.59 mSv, 85% is from natural sources and 15% from man-made sources; of that 15%, the major proportion is from exposures for medical purposes (a chest X-ray, for instance, would involve an effective dose equivalent of 0.02 mSv). There can be large variations in individual doses, particularly from natural radon exposure indoors, which gives the largest single contribution to the overall dose.

The ICRP have published recommended dose limits for exposures arising from human activities, other than medical exposures, both for the occupational

Table 19.7 *Average annual effective radiation dose received by individuals in the UK from all sources*

Source	Annual effective dose (mSv)	Range (mSv)	Total (mSv)
Natural sources			
Cosmic radiation	0.26	0.2–0.3	
Terrestrial gamma rays	0.35	0.1–1.0	
Radon decay products	1.30	0.4–100	
Other internal radiation	0.30	0.1–1.0	
		1.0–100	2.21
Artificial origin			
Medical procedures	0.370		
Weapons fall-out	0.005		
Discharges to the environment	0.0004		
Occupational exposure	0.007		
Miscellaneous sources	0.0004		0.38
TOTAL			2.59

workers and for the general public. For occupational workers, this is 20 mSv per year, and for the general public, 1 mSv in a year. These limits are based on the risk estimates for stochastic effects, and represent a level of dose above which the consequences for the individual would be regarded as unacceptable.

19.9 ROUTES OF EXPOSURE

Radiation exposure of an individual can be of two types: an external radiation exposure and/or an internal exposure.

19.9.1 External Exposures

These arise from sources of radiation outside the body that would normally be safely held within a suitable container. The dose received from such sources can be minimized by a combination of time, distance, and shielding:

- minimizing the time in which one is exposed to the source (dose = dose rate × time);
- maximizing the distance from the source (the inverse square law applies);
- using shielding (placing material between the source and the individual, which attenuates the radiation).

By a combination of these and with measurement of the dose received, the exposure can be kept to an acceptable level.

19.9.2 Internal Exposure

The situation can be more complex where there is an internal exposure. While this can arise from internalization of a radionuclide for medical purposes, when the amount taken will be known, it can also arise from inadvertent intake of material by ingestion, inhalation, through a wound in the skin, or contamination of the skin itself. Since this situation usually arises from inadvertent release of radioactive material, referred to as contamination, it can be difficult to determine the amount of radionuclide absorbed.

Once ingested, the radionuclide will be in intimate contact with the cells of the body, and will continue to irradiate tissues internally until either the radioactivity has decayed or it is excreted. The rate of excretion will depend on the chemical characteristics of the compound—its biological half-life (as distinct from the radioactive half-life). There is a particular hazard associated with radionuclides which possess long half-lives and that become fixed in particular tissues. An example of this is strontium-90, which is accumulated in the bone (being in the same Period of the Periodic Table as calcium) and in consequence is only slowly excreted, and which has a radioactive half-life of 29 years.

Because of this internal hazard, there is a need for secondary limits in terms of the amounts of individual radionuclides internalized, limiting the dose obtained from an internal radionuclide to the recommended dose limits discussed earlier.

The annual limit of intake (ALI) of a radionuclide is the amount of that radio-nuclide that would give a harm commitment to the organs it irradiates equal to that resulting from whole body irradiation of the annual effective dose limit. In determining the radiation burden, the committed effective dose [E(50)] is used. This is the dose from the radionuclide integrated over a period of 50 years, the likely maximum life span of the individual following the intake, and is of particular importance where the radionuclide is retained within the body for a long time.

Examples of ALIs of some radionuclides are given in Table 19.8. Apart from depending on the individual radionuclide, it can also vary according to the mode of entry into the body, whether by inhalation or ingestion, and on the fraction of inhaled or ingested activity that is absorbed (f_1). Clearance of inhaled material from the body may depend on its chemical form. Compounds of radionuclides are therefore classified into three groups depending on how rapidly they are cleared from the lungs: Class D, retained for days (*e.g.* K, Na, Ca); Class W, retained for weeks (*e.g.* Am, Cm); and Class Y, retained for years (*e.g.* PuO_2). The ALI will depend on the category to which the compound of the radionuclide belongs.

The values of ALI can also be used to set derived limits of concentration for radionuclides in air or water, given particular volumes inhaled or ingested in a year.

Table 19.8 *Some annual limits of intake (ALIs)*

Radionuclide			Inhalation			Ingestion	
	$t_{1/2}$	Class	f_1	ALI (kBq)		f_1	ALI (kBq)
^{42}K	12.36 h	D	1.0	50 000		1.0	50 000
^{137}Cs	30.0 y	D	1.0	2000		1.0	1000
^{45}Ca	163.0 d	W	0.3	10 000		0.3	20 000
^{90}Sr	29.12 y	D	0.3	400		0.3	600
		Y	0.01	60		0.01	5000
^{238}Pu	87.74 y	W	0.001	0.3		0.001	40
		Y	0.00001	0.3		0.0001	300
^{239}Pu	24065 y	W	0.001	0.3		0.001	40
		Y	0.00001	0.3		0.0001	300

19.10 METABOLISM OF RADIONUCLIDES

If a radionuclide absorbed into the body is an isotope of an element normally present (*e.g.* Na, K, or Cl), it will behave like the stable element. Also, if it has similar chemical properties to an element normally present, it will tend to follow the metabolic pathways of the natural metabolite (*e.g.* ^{137}Cs and K, or ^{90}Sr and Ca). For other radionuclides, their metabolism will depend on their affinity for biological ligands and for membrane transport systems. 'Labelled' organic compounds will, of course, follow the metabolic path of the corresponding 'cold' compound, and these will not be discussed further here.

After absorption into the body, the different inorganic elements can be classified into three groups according to their distribution in the body:

- elements that distribute throughout the body tissues;
- elements that concentrate in a particular organ or tissue;
- elements that concentrate in a number of tissues.

19.10.1 Elements that Distribute Throughout the Body Tissues

Examples of these are ^{42}K and ^{137}Cs. Both are readily absorbed from the gut (f_1=1.0), and because of their similar chemistry both will follow the behaviour of body potassium, although with some expected differences in quantitative rates of transfer across cell membranes. Both are excreted in the urine and the faeces.

19.10.2 Elements that Concentrate in a Particular Organ or Tissue

Calcium has an important function as a major component of bone, although bone also acts as a reservoir of calcium in the body: in man about 17% of calcium in the skeleton is recycled each year. About 30% of ^{45}Ca is absorbed from the gut (f_1=0.3) and around 65% of that is deposited in the skeleton. ^{90}Sr follows a similar route, although urinary excretion is greater.

The thyroid gland normally concentrates iodine to form an iodinated thyroid hormone precursor. ^{131}I therefore follows this pathway: it is readily absorbed from the gut (f_1=1.0) and from the lungs if inhaled. Although the major part is rapidly excreted in the urine, about 30% is accumulated by the thyroid, later being lost from the body with a $t_{1/2}$ of about 100 days. The proportion taken up by the thyroid decreases if the intake of stable iodide is increased.

19.10.3 Elements that Concentrate in a Number of Tissues

Plutonium will be taken as an example in this group. Being an alpha-emitter, it has a particularly high potential for damage to tissues. Its compounds can range from being soluble in water (*e.g.* plutonium nitrate or chloride) to being chemically inert and insoluble (plutonium dioxide). The soluble compounds readily hydrolyse in water at near-neutral pH forming an insoluble hydrated oxide. They can also complex with other molecules *in vivo*, thus remaining in solution.

Plutonium mainly enters the body by inhalation. The soluble component is rapidly absorbed from the lungs and transported in the blood to be either excreted through the kidneys or deposited in tissues (mainly bone and liver). The material remaining is in the form of the hydrated oxide polymer or insoluble compounds, such as PuO_2. This particulate material may be ingested by macrophages, which may then migrate to lymph nodes, or be removed by ciliary action, to be swallowed and excreted in the faeces. Some dissolution in lung fluids with transfer to the blood circulation may also occur.

Of plutonium activity entering the blood, about 45% is deposited in the liver,

45% in the skeleton, and the remainder either excreted or deposited in other tissues. Biological half-lives in the bone and liver are about 100 years and 40 years respectively. From animal studies it would appear that the lungs, the cells of the inner surface of bone, the bone marrow, and the liver are most at risk from accidental intakes of plutonium.

19.11 BIBLIOGRAPHY

K.J. Connor and I.S. McLintock, 'Radiation Protection Handbook for Laboratory Workers', HHSC Handbook No. 14. H and H Scientific Consultants, Leeds, 1994.

J.S. Hughes and M.C. O'Riordan, 'Radiation exposure of the UK population – 1993 Review', National Radiation Protection Board (NRPB) Report NRPB-R263. HMSO, London, 1993.

International Commission on Radiological Protection (1991), 1990 Recommendations of the International Commission on Radiological Protection, *Ann. ICRP*, 1991, **21**(1-3), 1–197 (ICRP Publication 60).

International Commission on Radiological Protection (1991), Annual Limits on Intake of Radionuclides by Workers based on the 1990 Recommendations. *Ann. ICRP*, 1991, **21**(4), 1–41 (ICRP Publication 61).

International Commission on Radiological Protection (1979), Limits for Intakes of Radionuclides by Workers. *Ann. ICRP*, 1979, **2**(3-4), 1–116 (ICRP Publication 30).

H.G. Paretzke, J.A. Dennis, and J.P. Massue, eds., 'Ionising Radiation and Protection of Man', A Book of Teaching Modules prepared by a European Expert Group. GSF Press, Neuherberg, 1981.

C. Keller, 'Radiochemistry'. Ellis Horwood, Chichester, 1988.

A. Martin, and S.A. Harbison, 'An Introduction to Radiation Protection'. Chapman and Hall, London, 3rd edn., 1986.

National Radiation Protection Board, 'Living with Radiation'. HMSO, London, 1989.

J. Shapiro, 'Radiation Protection. A Guide for Scientists and Physicians', Harvard University Press, Cambridge, MA, 3rd edn, 1990.

CHAPTER 20

Biocides and Pesticides

B. HEINZOW

20.1 INTRODUCTION

About 75% of the animal species in the world are insects. Some are beneficial predators and pollinators but many are pests acting as competitors for food. Others are vectors of infectious and parasitic disease. Not surprisingly, man has always tried to control such pests. In the past, preparations containing sulphur, arsenic compounds, extracts of tobacco and chrysanthemum, and strychnine were used, but only the synthetic pesticides produced by the application of modern chemistry have been really successful. Agricultural yield has increased dramatically over the last 50 years and biocides have played a major role in this. Unfortunately many of the compounds used may be harmful to the environment when used carelessly.

A biocide is any substance used with the intention of killing living organisms. A pesticide is defined as any substance intended for killing pests. The ideal pesticide is one that is toxic primarily to the target pests and is rapidly inactivated in the environment, but few such pesticides exist. The development of more selective, less persistent, and safer pesticides is one of the great demands on chemistry of the future.

Biocides and pesticides may be more specifically classified into the following groups, most of which are defined in the glossary at the end of this book.

Acaricides
Arachnicides
Avicides
Fungicides
Herbicides
Insecticides
Larvicides
Miticides
Molluscicides
Nemat(o)icides
Pediculicides

Piscicides
Rodenticides
Scabicides

Although this grouping suggests selectivity of the types of biocide listed, it must be emphasized that biocides may be harmful to any living organism including man depending only upon the route of exposure and dose.

There is, at present, great concern about pesticide residues in food and in the environment and the possibility of harm to humans following long-term low-level exposure and consequent effects of chronic poisoning. These problems will not be considered here, but it should be remembered that a risk may be tolerable if sufficient benefit is obtained from use of a biocide. For example, a small risk from chronic exposure may be accepted from a biocide which helps to eradicate diseases like malaria and river blindness.

Table 20.1 shows some of the effects on humans that have been associated with excessive exposures. These reflect the target organ toxicity of pesticides.

Table 20.1 *Toxic effects of biocides on humans*

Compound	Symptoms of intoxication
Aniline	Methaemoglobinaemia, irritation of skin and mucous epithelium, sensitization
Chlorinated hydrocarbons	Central nervous system: neurotoxicity, seizures, tremor, hypotension
Dinitrocresol	Thirst, sweating, hyperthermia, vertigo, vomiting, diarrhoea, cyanosis, arrhythmias, pulmonary oedema, liver and kidney failure
Dithiocarbamate	Irritation of skin and mucous epithelium, vertigo, vomiting, diarrhoea
Phenoxycarbonic acid	Irritation of skin and mucous epithelium (ulceration), muscle fibrillation, peripheral neuropathy (pain), hyperglycaemia
Carbonic acid	Irritation of skin and mucous epithelium, ulceration, and necrosis
Dipyridinium	Irritation of skin and mucous epithelium, vomiting, diarrhoea, hyperthermia. After latency, renal failure, liver damage
Paraquat	Lung fibrosis after latency period
Organophosphate esters	Headache, blurred vision, sweating, dyspnoea, vomiting, cyanosis, muscle fibrillation, seizures, miosis, bradycardia
Carbamate	Similar symptoms to organophosphate esters, faster onset and decline
Pyrethrum, Pyrethrin	Central and peripheral neurotoxicity, irritation, paraesthesia, seizures, sensitization
Pyrethroids	Tremor, salivation, chorea, paraesthesis
Organotin	Central nervous system: neurotoxicity, depression, anorexia, diarrhoea, headache, vertigo, vomiting, blurred vision
Coumarin	Blood clotting disturbance

20.2 ORGANOCHLORINE INSECTICIDES

Dichlorodiphenyltrichloroethane (DDT) was the first of a variety of contact organochlorine (chlorinated hydrocarbon) insecticides which include aldrin, dieldrin, endrin, chlordane, and hexachlorobenzene. These compounds were used extensively in the mid-1940s to 1960s. Their properties of low volatility, chemical stability, and environmental persistence led to their bioaccumulation (bioconcentration, biomagnification) in the food chain of fish, birds, and mammals owing to their lipophilicity and slow metabolic degradation. It was then demonstrated that these compounds possess strong estrogenic and enzyme-inducing properties which interfere with the reproductive system, although the presenting symptoms are varied and non-specific (Table 20.2) In avian species of a high trophic level like pelicans, seagulls, and eagles, the adverse effects of the dichlorodiphenyl derivatives is related to induction of steroid-metabolizing enzymes and the inability of the reproductive organs to mobilize enough calcium in the production of the eggshell. This eggshell-thinning leads to cracks allowing bacteria to infiltrate with resultant death of the fetus even if there is not complete breakage of the egg in the nest.

Table 20.2 *Symptoms of organochlorine poisoning*

Compound/group	Acute symptoms	Chronic symptoms
Dichlorodiphenyls	Paraesthesia, ataxia, dizziness, headache, nausea, vomiting, fatigue, lethargy, tremor	Anorexia, weight loss, anaemia, tremor, weakness, hyperexcitability, anxiety
HCH, Lindane	Headache, tremor, convulsion	
Cylodienes	Dizziness, headache, nausea, vomiting, hyperexcitability, hyper-reflexia, myclonic jerking, convulsion	Dizziness, headache, hyperexcitability, muscle twitching, myoclonic jerking, anxiety, insomnia, irritability, epileptiform convulsions
Chlordecone, Mirex		Arthralgia, rash, ataxia, slurred speech, blurred vision, irritability, loss of memory, weakness, tremor, decrease of sperm count

Recently the hormone-like activity of dichlorodiphenyl dichloroethylene has raised concern in connection with the steady decline of sperm counts in man over the past 30 years. One interesting characteristic of the physicochemistry of these compounds is the extent of partitioning determined by Henry's law constant. Depending on their low or intermediate vapour pressure, compounds like DDT and toxaphenes volatize in hot climate zones and as the temperature drops condense in polar regions by partitioning into particles, water, and aerosols. This global distillation explains the transfer to, and accumulation in, wildlife species of colder areas.

20.2.1 Mechanism of Toxic Action of Organochlorines

DDT-type insecticides interact with the neuronal membrane by altering the membrane permeability (transport) for potassium and sodium and the calcium-mediated processes. By inhibiting these functions, the repolarization of the nerves is disturbed resulting in hyperexcitability.

Cyclodienes and cyclohexane compounds have a central nervous system-stimulating mode of action. These compounds antagonize the neurotransmitter, γ-aminobutyric acid (GABA), permitting only partial repolarization of the neuron and, thus, uncoordinated nervous excitation.

Owing to their lipophilicity, organochlorine compounds are partitioned and stored largely in the adipose tissue, where they are biologically inactive. There is an equilibrium between body fat and free circulating compounds. Redistribution and mobilization of fat, for example due to disease, ageing, or fasting, may result in mobilization of stored organochlorines in quantities that may lead to manifest toxicity. Owing to the different patterns of use of organochlorine pesticides in industrialized and developing countries, the pattern of distribution of the residues in human fat (including breast milk) differs between countries.

20.3 ORGANOPHOSPHATES AND CARBAMATES

A large number (>200) of insecticides are derived from esters of phosphoric acid, phosphorothioic acid, and carbamic acid. These potent insecticides produce their biological action by blocking the nervous tissue enzyme acetylcholinesterase (AChE) in ganglia and in the parasympathetic nervous system (Figures 20.1 and 20.2). This enzyme hydrolyses the neurotransmitter acetylcholine (ACh) and, if it is inhibited, ACh accumulates at the nerve endings (Figure 20.3.) causing

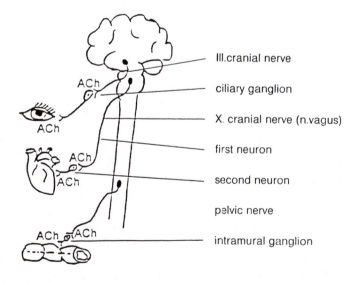

Figure 20.1 *The parasympathetic autonomic nervous system*

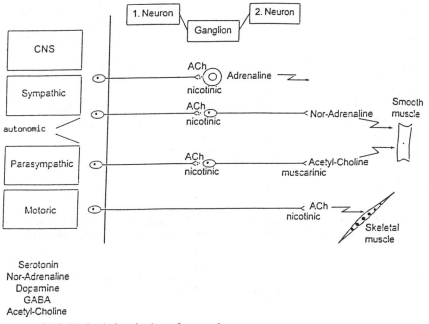

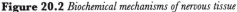

Figure 20.2 *Biochemical mechanisms of nervous tissue*

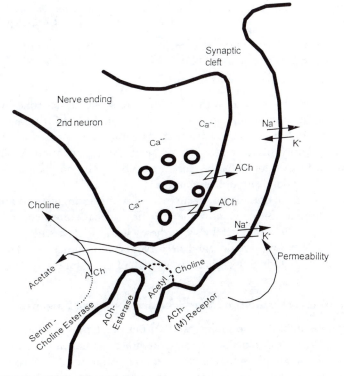

Figure 20.3 *Inhibition of acetylcholinesterase*

continual and uncoordinated ACh stimulation of the muscarinic receptors of the parasympathetic nervous system, and the stimulation and blocking of nicotinic receptors in the ganglia of the autonomic nervous system, the motor muscles, and the central nervous system.

Two further manifestations of toxicity may occur with some compounds:

(i) intermediate syndrome, a paralytic condition mainly of cranial nerves (caused by fenthion, dimethoate, monocrotophos, methamidophos);
(ii) organophosphate-induced delayed neurotoxicity (caused by leptophos), with symptoms of pyramidal tract damage such as spasticity.

(See Table 20.3.)

Table 20.3 *Symptoms of organophosphate and carbamate inhibition of acetylcholine esterase*

Acute symptoms	Chronic symptoms
Miosis, hypersecretion, bronchoconstriction, diarrhoea, cramps, urination, bradycardia, cardiac arrest, muscle fasciculation, tremor, weakness, paralysis, restlessness, ataxia, lethargy, confusion, loss of memory, convulsion, respiratory depression, coma	Loss of ambition and libido, autonomic dysfunction, cephalgia, gastrointestinal symptoms Delayed and lasting neuropathy and neuropsychological dysfunction, emotional lability, confusion, memory loss, anxiety, depression, insomnia, ataxia, speech difficulty, muscle weakness, neurological deficits

20.3.1 Mechanism and Site of Action

Organosphosphorus esters react with the active site of AChE (a serine hydroxyl group) resulting in a phosphorylated and inhibited enzyme. With some insecticides (and particularly the related war nerve gases) this process is irreversible and duration and severity of toxicity are prolonged. New-generation organophophates have been developed with more reversible phosphorylation allowing spontaneous dissociation from the nervous tissue.

Carbamate esters are bound to the active site of AChE forming a carbamylated and inhibited protein. Subsequently, the enzyme is decarbamylated and active again. In principle the difference between these chemical classes lies in the rate constant of the dephosphorylation and decarbamoylation processes, the former being orders of magnitude slower than dissociation of the natural substrate ACh.

20.3.2 Specific Treatment of Acetylcholine Esterase Poisoning

Choline esterase inhibitor poisoning is a serious, and possibly fatal, medical emergency and requires immediate treatment with the antagonist atropine and, possible only for organophosphate poisoning, reactivation of AChE with oxime therapy. The degree and time course of the intoxication can be monitored by analysis of the activity of non-specific serum choline esterase and of specific erythrocyte AChE. The life-threatening symptoms of respiratory depression and hypoxia require immediate, adequate, medical and supportive intervention.

20.4 PYRETHROID INSECTICIDES

Synthetic pyrethroids are currently the most widely used pesticides. Historically they were extracted from chrysanthemum flowers which contain the insecticidal esters (pyrethrins). Being very potent insecticides with low mammalian toxicity they have gained a strong market share. Based on the characteristics of poisoning, two classes of pyrethroids are distinguished, as shown in Table 20.4.

Table 20.4 *The two classes of pyrethroids*

Chemical characteristics	Toxicity class	Symptoms
Class 1 Pyrethrin, allethrin, tetramethrin, resmethrin, phenothrin, permethrin	T-syndrome Action on the central and peripheral nervous systems	Hyperexcitation, tremor, prostration
Class 2 α -cyano substituted: cypermethrin, deltamethrin	CS-syndrome Action on the mammalian central nervous system	Cutaneous paraesthesia, salivation, chronic seizures, dermal tingling

With natural pyrethrum, allergic sensitization occurs but rarely any other toxic effect. Inappropriate use and careless handling of pesticides has resulted in some cases of human intoxication. Most prominent symptoms are burning, itching, tingling sensations of the skin, paraesthesia upon dermal contact and, after ingestion, epigastric pain, nausea, vomiting, headache, vertigo, blurred vision, muscular fasciculations, convulsions, and seizures.

20.4.1 Mechanism of Action

The type I esters affect the sodium channels in nerve membranes with prolongation of sodium influx causing repetitive neuronal discharge and prolonged afterpotential but no severe membrane depolarization. The type II α-cyanoesters lead to greater and more prolonged sodium influx with persistent membrane depolarization and eventually nerve blockade. Whereas the first group exerts its main effects on synaptic transmission causing hyperexcitability and tremor, the second group shows its first effects on the sensory nervous system. Other actions include inhibition of Na^+/Mg^{2+}ATPase (adenosine triphosphatase) and alteration of calcium and chloride ion homeostasis.

20.5   REACTIVE DERIVATIVES OF PESTICIDES

Biocides like other xenobiotics are subject to metabolic alteration, for example by the phase I and II reactions. The derivatives produced are usually less biologically active (detoxification), but sometimes are more toxic (biotoxication). The herbicide, paraquat, is known to cause lung damage in mammals as a result of free radical formation and resultant modification of membrane lipids. The free radicals are produced during the reduction of paraquat by nicotinamide adenine dinucleotide phosphate (NADP)–cytochrome P_{450} reductase. Phosphorothioate insecticides are bioactivated through oxidative desulfuration. Parathion is converted by cytochrome P_{450} monooxygenase to paraoxon which exhibits anticholinesterase activity. Other examples of metabolic conversions by cytochrome P_{450} monooxygenases include S-oxidation of disulfoton, fenthion, and di-allate, the hydrolysis of butonase, trichlorphon, and propanil, the dehalogenation of naled, tralomethrin, and tralocythrine (the brominated derivatives of delta- and cypermethrin), and the formation of lipophilic conjugates of fenvalerate, which cause granulomas in severals tissues of rats and mice.

A good example of synergistic potentiation occurs with non-toxic exposure to chlordecone which produces a nearly 70-fold amplification of chloroform lethality. The increase in chloroform toxicity is related to the arrest of hepatocellular regeneration by chlordecone. Thus, hepatocellular injury by a normally nonlethal exposure to chloroform progresses from activated lipid peroxidation into lethal liver necrosis.

20.6   PESTICIDE RESIDUES IN FOOD AND DRINKING WATER

Pesticide residues in food and drinking water must be considered in any adequate risk assessment of the effects of the use of pesticides. Pesticides are currently perceived by the public as posing a major health threat to the well-being of the population. In consumer surveys usually about 80% of those questioned consider pesticide residues as a major concern.

Processes of risk assessment for pesticide residues are based on three components: estimation of residues in food, estimation of food consumption patterns from national surveys, and toxicological characterization by comparison of exposure estimates with data from animal toxicology studies. Commonly the theoretical maximum legal residue level is used in the calculation. For extrapolation from animal experiments to man and the calculation of acceptable daily intake (ADI) of the United Nations Food and Agriculture Organisation/World Health Organisation, or the reference dose (RfD) of the US–Environment Protection Agency, two main assumptions are made: (i) effects in animals can be used to predict effects in man and (ii) effects of high doses can be mathematically related to effects (or lack) of low doses. To accommodate the uncertainty, safety (or uncertainty) factors are often applied. These range from 10 to 10 000, 100 being the most commonly used. If exposure is below the ADI or RfD (the amount of a substance a human can consume daily without harm) the risk associated is considered insignificant.

In most cases the maximum residue concentrations found in food are far below the ADI and RfDs. However, owing to misuse or inadvertent addition of pesticides to food, cases of intoxication may occur. For example the illegal application of aldicarb to watermelons resulted in several hundred cases of acute cholinergic intoxication in western USA in 1985.

The most commonly involved pesticides in human poisoning are organophosphates, carbamates, chlorinated hydrocarbons, and organic mercurials. Food is the most common vehicle of exposure, followed by skin contact, and respiratory exposure. Pesticides have made their contribution to the quantity and quality of food available and their use will continue for many years, but there is a need for monitoring environmental impact and long-term health effects in man of low-level exposure to ensure that the benefits of pesticides can be enjoyed without any accompanying harm.

20.7   CONCLUSIONS

Pesticides are generally designed to kill higher organisms and thus pose special toxicological problems. However, despite the development of some non-chemical methods of pest control, there is at present no generally satisfactory alternative to chemical control. Between total ban and careless use, there is a pathway of safe, careful application following rational risk–benefit asssessment and this is the aim of appropriate pesticide management, *i.e.* to optimize the benefits obtained while minimizing the risks.

20.8   BIBLIOGRAPHY

L.G. Costa, C.L. Galli, and S.D. Murphy 'Toxicology of Pesticides: Experimental, Clinical and Regulatory Perspectives', NATO ASI Series H: Cell Biology, Volume H13. Springer, Berlin, 1987.

DFG (German Research Foundation) 'Okotoxikologie von Pflanzenschutzmitteln'. Verlag Chemie VCH, Weinheim, 1994.

H. Frehse ed. 'Pesticide Chemistry',Verlag Chemie VCH, Weinheim, 1991.

C.D. Klaassen, ed. 'Casarett and Doull's Toxicology'. McGraw-Hill, New York, 5th edn, 1996.

K.T. Maddy, S. Edminston, and D. Richmond 'Illnesses, injuries and death from pesticide exposures in California 1949–1988', *Rev. Environ. Contam. Toxicol.*, 1990, **114**, 57–123.

R.E. Gosselin, R.P. Smith, and H.C. Hodge 'Clinical Toxicology of Commercial Products'. Williams and Wilkins, Baltimore, 5th edn, 1984.

C. Tomlin 'The Pesticide Manual'. British Crop Protection Council/The Royal Society of Chemistry, 10th edn, 1994.

P. de Vroogt and B. Jansson 'Vertical and long range transport of persistent organics in the atmosphere', *Rev. Environ. Contam. Toxicol.*, 1993, **132** (complete volume).

G.W. Ware and H.N. Nigg eds. 'Minimizing human exposure to pesticides',

Symposium proceedings, American Chemical Society, San Francisco 1992, *Rev. Environ. Contam. Toxicol.*, *1992*, **128–9** (two complete volumes).

R.D. Wauchope, T.M. Butler, A.G. Hornsby, P.W.M.A. Beckers and J.P. Burt 'The SCS/ARS/CES Pesticide Properties Database for environmental decision-making', *Rev. Environ. Contam. Toxicol.*, 1992, **123**, (complete volume).

C.K. Winter 'Dietary pesticide risk assessment', *Rev. Environ. Contam. Toxicol.*, 1992, **127**, 23–67.

CHAPTER 21

Safe Handling of Chemicals

HOWARD G.J. WORTH

21.1   INTRODUCTION

The introduction to this monograph states that toxicology is the fundamental science of poisons. In the broadest sense of this definition the adverse action of any chemical on living tissue may be toxicological. This ranges from the corrosive effect of the spillage of a strong mineral acid, to the injurious effect of exposure to radionuclides, to the exposure to biological fluids containing potentially dangerous pathogens, and to environmental and occupational exposure to chemicals. The chemist is often seen as the person who can give help and guidance on the handling of chemicals, on the toxicological effects associated with them, and advice on how to deal with an incident that occurs. It is not surprising that chemists are regarded as the professional group that should be able to give this help and advice. However, this is frequently not recognized in the curriculum for the training of chemists and indeed, apart from what they pick up as part of their educational progress, there is often no formal teaching of toxicology. This makes chemists vulnerable as there is considerable legislation in many countries concerned with the toxicity of the handling of chemicals.

21.2   LEGISLATION

In recent years, many countries have passed legislation concerned with safe practice in the working place. Inevitably, this has implications for the handling of chemicals. The considerable use of chemicals in the domestic and non-technical environment means that their safe handling is no longer just a concern of those employed in the chemical industry. Domestic cleaners, solvents and detergents, weed killers and pesticides, and proprietary medicines are examples of chemicals available to the public, whose safe handling may well be the subject of legislation or other related documentation.

In the UK in 1974 the Health and Safety at Work Act was passed which required the appointment of safety representatives, the establishment of safety committees, and the inspection of the workplace by representatives and/or members of the safety committee. This has implications for handling of chemicals where these are used in the workplace. The Act also indicated that every employer is responsible for ensuring, through his or her actions, that the workplace is a safe environment. This has implications for the most junior chemistry graduate in a laboratory who will be held responsible, by the Act, for his or her safe use and handling of chemicals.

As a consequence, documentation has been produced concerned with specific aspects of handling potentially toxic materials, for example the Department of Health and Social Security produced a series of documents culminating in 'Code of Practice for the Prevention of Infection in Clinical Laboratories and Post Mortem Rooms' in 1978. This has now been superseded by 'Safe Working and the Prevention of Infection in Clinical Laboratories' produced by the Health and Safety Commission. More recently, regulations for the Control of Substances Hazardous to Health (COSHH), require laboratories and institutions to compile and retain inventories of chemicals used. This must include statements concerning their toxicity, safe handling procedures, and action that should be taken in the event of an accident or spillage. Guidelines are now produced by many chemical reagent supply companies which are invaluable compendia of the hazards related to individual chemicals.

Such legislation is not restricted to the UK; most European and North American countries have similar requirements, which will undoubtedly be adopted by developing countries. For example, the International Programme on Chemical Safety, which is a division of the World Health Organization is now producing international chemical safety cards in conjunction with the Commission of the European Communities.

21.3   TOXICOLOGICAL REACTION

It is impossible to deal with every conceivable chemical toxicological reaction in a monograph such as this, and indeed some areas are dealt with in more detail in other chapters. This chapter will highlight some of the more common areas of possible reaction and give examples of these.

21.3.1 Corrosion

Corrosive chemicals have a wholly destructive action on tissues. This is often a hydrolytic reaction through the high water content of the cell leading to structural and chemical destruction, which manifests itself in burns of varying degrees of severity according to penetration and the degree of contact. This occurs most commonly through skin reaction but not exclusively so, as corrosive chemicals could be ingested, or even inhaled. An obvious example is the reaction of a strong mineral acid or base. Acid salts such as phosphorus(v) chloride are similarly reactive but neutral salts are not. Contact with sodium chloride is not injurious, although it is corrosive in the sense that it will react with other compounds, metals for example. But the ingestion of large quantities of sodium chloride would be toxic because of its absorbance into cells, which would not occur by normal skin contact. The toxic effect of corrosive chemicals is not confined to inorganic compounds, although corrosive organic compounds would usually have a relative low or high pK value as with ethanoic acid or tertiary ammonium compounds, or have an inorganic component as in benzoyl chloride.

21.3.2 Organic Compounds

The range of compounds that might be considered under this heading is enormous and wide ranging. When toxicity arises primarily due to the organic nature of the compound, it is often a consequence of its volatility. With this may go the descriptions of solvent and flammability. These terms are often used to indicate toxicity, but are misleading in that there are many toxic organic compounds that are not solvents, and some organic solvents that are toxic, but not flammable. Secondly, flammability clearly indicates a hazard but is not necessarily an indication of toxicity. The confusion probably arises because most organic solvents are flammable and toxic. Almost by definition they are good solvents because they are volatile, and because they are volatile they are easily absorbed, through the skin and by inhalation. However, the student needs to be clear about these differences, and they are best demonstrated by example.

Benzene derivatives and ketones form the basis of many industrial solvent mixtures, used for example in adhesive preparations whose misuse is frequently referred to as 'glue sniffing'. Their toxicity is well recognized through inhalation and they are good solvents, volatile and flammable. The simple alcohols are good solvents, they are flammable and toxic but their toxicity is due to ingestion, not inhalation. Other solvents are volatile and toxic but not flammable, for example the halogenated hydrocarbons such as chloroform and carbon tetrachloride. Carbon tetrachloride is sufficiently toxic, by inhalation, that it has been largely withdrawn as a commercially available chemical and as a domestic dry cleaning agent. Similarly, heterocyclics, pyridine and its derivatives, are good solvents, volatile and highly toxic, but non-flammable.

The modes of action of these various groups of compounds may be very different, the alcohols have a long-term chronic, toxic effect on liver cells, whereas halogenated hydrocarbons, which are also extremely hepatotoxic, cause acute liver failure. Others such as benzene and related derivatives, and aromatic amines have carcinogenic properties; this makes them toxic by ingestion as well as inhalation. It is interesting to note that toluene-based compounds are much less toxic in this respect than benzene derivatives, and toluene is therefore preferred as a solvent. This is because it is metabolized to benzoic acid which is water-soluble and may therefore be excreted through the kidneys. It is, nonetheless, highly flammable.

Finally, extreme examples of rapid toxicity are those that directly affect the central nervous system or the mitochondrial respiratory pathway, for example many organophosphorus compounds and cyanides.

21.3.3 Biological

Chemistry students frequently study modules in biochemistry and other biological sciences, and may later embark on a career in biological sciences. This involves the handling of biological fluids and substances, and the risk of exposure to dangerous pathogens. In the handling of human material the most common, potentially serious pathogens are the hepatitis viruses and human immunodeficiency virus (HIV). The pathological activity of body fluids and other materials is dependent upon the likelihood of containing the virus, but all should be treated as potentially hazardous. The most likely hazard to laboratory workers handling pathogens is through introduction of the fluid into the blood stream, which can occur through contact with cuts, abrasions or other breakages in the skin's surface or through needle stick injuries. The likelihood is extremely low, but the risk is high because of the prognosis of contact with certain pathogens, the most notable being HIV. For this reason the chemist must know how to handle biological fluids safely and be aware of the appropriate use of protective clothing.

21.3.4 Allergens

Individuals may have an allergic reaction to compounds that are otherwise harmless. This involves stimulation of the immune system to produce an apparently unnecessary defensive mechanism. It may occur as a result of sensitization to a particular compound due to an earlier exposure. The possible reaction, both in terms of the allergen and the seriousness of the reaction, is enormously variable. Almost any compound can be responsible, for example, adverse reactions are well recognized with diazomethane, formaldehyde, isocyanates, phenols, nickel salts, and chromates(vi).

Common allergies are also recognized in certain foodstuffs, strawberries for instance, presumably due to the presence of a particular compound, and in pharmaceutical preparations such as penicillin. The reaction is widespread, ranging from an irritating, benign rash, to a life-threatening anaphylactic reaction. Some individuals are much more easily sensitized than others and it is therefore impossible to lay down any specific requirements for the safe handling of chemicals that may cause an allergic reaction. If good handling procedures are operated, the risk is minimal. However, if an individual is aware of being sensitive to a particular compound then it should be handled with special care or its use avoided completely.

21.3.5 Pharmaceuticals

Many chemists, after graduation, will find themselves working in the pharmaceutical industry and consequently handling chemicals with a pharmacological activity. Most pharmaceuticals are not toxic as such and will therefore not cause any hazard unless ingested in quantities which indicate malicious rather than accidental ingestion or exposure. Consequently, if safe chemical handling procedures are adopted, pharmaceutical compounds should not require special mention.

However, there is a security aspect to the safe handling of pharmaceutical compounds that have an unlawful market. The well-known 'hard line' drugs such as opiates clearly fall into this category, as may others that are not available on the open market, and particularly those that are no longer readily available through medical practitioners, for example barbiturates and benzodiazepines. High security cupboards are therefore recommended for the storage of such compounds. In many countries legislation requires such security for certain prescribed drugs.

21.3.6 Radionuclides

The use of radionuclides has been discouraged over recent years on health and safety grounds, and in many instances alternative methods have been introduced where appropriate. This is particularly true in the analytical field of immunoassay where radioactive isotopes are being replaced by alternative labels. Nonetheless, radionuclides are still used and handled by chemists in many different areas. Their use and handling is controlled by legislation in most countries, and this is often complex and variable because the hazards involved are dependent upon the type of emission and nature of the labelled compound. For example an isotope in a compound that may be metabolized if ingested, can end up in many other compounds in the body, in some instances concentrated in particular organs, which constitutes a much greater hazard than the ingestion of the same isotope in a compound that is not metabolized. For these reasons legislation requires purpose-designed laboratory areas put aside solely for the handling of isotopes, and the safety level of this area may be graded according to the types of isotopes and compounds that are going to be handled.

For further details, reference must be made to local legislation in which the requirements are indicated. This may also require inspection by government inspectors before the handling of radionuclides is permitted.

21.4 GOOD LABORATORY PRACTICES

The chemist who is trained in good laboratory practice will be practising the safe handling of chemicals. It is impossible to lay down criteria that will guarantee safe handling because it is not possible to predict the eventualities and situations that could arise in a chemical laboratory. However, safe practice, whether it be in a chemistry laboratory or any other workplace, consists of a combination of common sense, experience, and a technical understanding and appreciation of the procedures that are being carried out.

21.4.1 Storage

A well-organized laboratory will have a continuous supply of the chemicals that are required for its work, which means there must be a comprehensive storage system. The number and quantities of chemicals retained at the bench should be minimal, the bulk being kept in a controlled and safe store. Bench chemicals should never be stored on the actual bench, but on shelving where they are easily accessible without the operator having to stretch unduly to reach them or to stretch over equipment on the bench, thus running the risk of protective clothing being caught in moving apparatus or coming into contact with chemicals in use on the bench.

Commercially purchased chemicals should be stored in the bottles or containers in which they are received from the manufacturer, thus preserving the correct manufacturer's labelling. Reagents prepared within the laboratory should be clearly labelled and contained in an appropriate receptacle. There should never be more than one container of a given chemical or reagent stored on the working bench at any time.

Where flammable organic solvents are used, not more than 500 ml should be stored at the working bench. Compounds that are particularly hazardous should be stored appropriately. For example, working quantities of flammable solvents should be stored in a steel storage cabinet, not on the bench shelves, and compounds known to have high toxicity should be stored in a fume cupboard. The fume cupboard should be designated for storage, and not used as a working cupboard.

Chemicals that are not intrinsically hazardous, but are highly toxic if ingested, need to be stored in a special high-security cabinet. This includes extreme, acute poisons, and some pharmaceutical drug preparations. Indeed, the handling of some substances may require government authority. In the UK for example, a Home Office licence is required for the storage and use of certain categories of drugs, in which case a high-security cabinet is required, and an inventory of the purchase and use of such compounds must be maintained.

It has already been mentioned that the bulk of chemicals should not be stored in the laboratory, but in an appropriate safe store. For most, a clean, dry store with convenient access and appropriate shelving is adequate. An inventory of the store must be maintained and a sequence of chemicals instituted. The obvious sequence is an alphabetical one, but this is not always recommended as it may result in the close proximity of storage of two or more chemicals which, if spilt, would result in a potentially dangerous reaction. In the absence of alternative satisfactory sequencing, alphabetical storage may be used, so long as it is overridden where known hazardous combinations might occur. The store area needs to be well ventilated and equipped with appropriate safety facilities, fire extinguishers, smoke detectors, drip trays, *etc.*

A special fire-proof store must be used for the storage of flammable solvents and strong acids. This must be a separate building, located near to, but outside, the laboratory complex and must comply with local safety regulation requirements. It must be able to maintain a major spillage and be equipped with appropriate safety facilities, *i.e.* smoke and fire detectors, fire extinguishers, suitable ventilation, *etc.*

All store areas must be specifically designated, and used for no other purpose.

21.4.2 Reagent Preparation

Laboratories buy many of their reagents already prepared from commercial suppliers, but some, particularly those used in a research environment, are prepared in the laboratory. It is therefore important that the student learns how to prepare reagents correctly and safely. In mixing chemicals, the student must be aware of possible reactions of mixing. These may be potentially hazardous as in the simple diluting of concentrated sulfuric acid, where sufficient heat may be generated to cause the resultant solution to boil. In this case the concentrated acid should be added to the water and not vice versa, thus if there is a spillage as a consequence of overheating, it is a dilute acid solution that is spilt, not a concentrated one. All bench techniques involved in reagent preparation must be carried out carefully to avoid chemical spillage whether it be on the balance pan during weighing or the bench or the floor during the transfer of chemicals. The reagent, when prepared, must be transferred to a suitable receptacle and labelled correctly. The labelling should include details of all constituent components and their quantities. The date upon which the reagent was prepared must be recorded as should the name of the person who prepared it.

Reagents must be prepared and stored in appropriate containers. In most cases this will be a glass bottle with a glass stopper. However, there are exceptions and these must be known and observed. For instance, solutions containing alkali at a pH greater than 12.0 should be stored in polythene bottles with inert plastic stoppers or screw tops. Glass bottles should not be used and under no circumstances glass bottles with glass stoppers. Strong hydroxides will react slowly with silicates in the glass and atmospheric carbon dioxide, and will form thin but hard layers of silicates and carbonates around the stopper making it extremely difficult to remove. There are more injuries in chemistry laboratories through breakage of glassware than any other single cause, and the removal of glass stoppers is probably the most common single cause of accidents with glassware.

Picric acid solution is another example of a possible hazard. The solution is stable, but the solid is potentially explosive. A thin layer of dried picric acid around the neck of a glass stoppered bottle may be quite dangerous. Organic solvents should not be stored in plastic bottles. Although many plastics such as polythene are inert to most organic solvents, there is a long-term effect which will cause the vessel to split suddenly. These are a few examples of many reagent vessel storage problems, which the student can only learn by experience.

Reagents that are light-sensitive need to be stored in dark glass bottles, and those that are very sensitive should be stored in dark glass bottles in a cupboard. Reagents that are labile at room temperature need to be stored in a refrigerator or even in a deep freezer. However, organic solvents must never be stored in a domestic-type refrigerator or freezer because of the risk of ignition through electrical discharge when the thermostat switches on or off. Only cooler units specifically designed to contain flammable materials should be used for such storage.

21.4.3 Biological Fluids

The potential danger of biological fluids has already been discussed from which it is clear that the hazard is dependent upon what, if any, infective agent is present. Every fluid should therefore be regarded as potentially infected and be handled accordingly. Good practice is largely a matter of hygiene and the appropriate use of protective clothing. These are dealt with below, under Health and Safety. In addition, there should be a laboratory policy for the swabbing of benches and other areas with a suitable disinfectant at frequent defined intervals. Under no circumstances should biological fluids be apportioned by mouth pipetting. Indeed mouth pipetting of any liquid should not be permitted. Pipette fillers should be used at all times and if they become contaminated by biological fluid, an appropriate decontamination procedure should be adopted, or, if necessary, the filler discarded.

21.4.4 Radionuclides

The hazard and possible toxic effects of radioisotopes are dependent upon the radiation emitted and the metabolism of the labelled compound if ingested. This has also been referred to earlier, as has the fact that many countries enforce strict legal requirements for the use and handling of isotopes. In addition to complying with these requirements, safe handling should be achieved by adhering to the same requirements as those for handling biological fluids.

21.4.5 Gases

Many laboratory procedures require the use of gases supplied commercially in pressurized bottles. These present two hazards, those associated with the chemical hazard of the gas, and those associated with a gas under pressure, although the gas itself may be chemically inert. It must be remembered that a serious leak of an inert gas can cause asphyxiation by replacement of the atmospheric oxygen.

A fully charged cylinder may have a pressure as high as 15 MPa, *i.e.* in excess of 100 atmospheres. The mishandling of gas cylinders therefore represents a considerable safety hazard. Cylinders should not be stored or operated from within the laboratory, but kept in a purpose-built store with appropriate protection, ventilation, and plumbing. If it is necessary to store a cylinder in the laboratory, it must be contained in a purpose-built stand and clamped securely to the bench. Transportation of cylinders should only be carried out with an appropriate purpose-designed cylinder carrier.

Cylinders should be operated from the store where the one in use is plumbed into the laboratory through a permanent piping system. Expert installers must be consulted in order to ensure the correct material is used for each gas (this is not necessarily the same for every gas). Reducing valves must be used at all times with high-pressure cylinders, and the operator must ensure the appropriate valve is used and beware of the differences between valves. Generally, flammable gases are fitted with left-hand thread valves whereas other gases are fitted with conventional right-hand thread valves.

Acetylene is probably the most explosive and dangerous chemical that is commonly used in chemical laboratories and there are a number of special precautions required in handling this gas in cylinders, both in relation to the cylinder and the plumbing. Chemists using acetylene must familiarize themselves with these requirements and satisfy themselves that they are being adhered to.

Hydrogen is almost as dangerous because of its explosive nature and because of its very low molecular mass. It is the smallest known molecule and as a consequence is highly diffusible. This enables it to 'creep' and accumulate in potentially explosive mixtures at sites some distance from the source if a leak occurs. This is a hazard of which the chemist must be well aware.

When cylinders are being moved, whether full or empty, they must be handled with care. They should not be rolled or manhandled, they should not be moved with the reducing valve in place because of the risk of damage, nor should the main valve be left open when the cylinder has been emptied. Finally, grease should never be used on any cylinder for lubrication unless it is specified for use on that cylinder. Flammable and non-flammable gases should be stored separately, and separate from oxygen.

21.4.6 Equipment

The safe use of equipment is related to the safe handling of chemicals in that if equipment is not used correctly then there is a greater likelihood of accidents with undue exposure to chemicals and their toxic and hazardous properties. Operational equipment should be maintained to a high level of electrical and mechanical safety, if necessary by having a service contract with the supplier. A sequence of check procedures must be established and adhered to. General laboratory tidiness is part of good equipment maintenance. Loose paper or reagent bottles or other small pieces of equipment left lying in inappropriate places may result in the blocking of air vents causing overheating or chemical spillage over equipment.

Finally, there are specific hazards related to particular pieces of equipment; for example, equipment that requires gas supplies from high-pressure bottles, some of which may be flammable and potentially explosive. Gas chromatographs not only require flammable gas supplies, but also have ovens set to relatively high temperatures. Centrifuges, because of their high speed of operation, pose a particular hazard. In many countries their use is now controlled by legislation, particularly if they are being used for processing biological fluids. These are examples of specific equipment hazards and the student needs to be familiar with the particular hazards associated with the equipment that he or she is using, and must adopt a good practice approach to all equipment.

21.5   HEALTH AND SAFETY

Good laboratory practice and the safe handling of chemicals is related to good health and safety practice which includes standards of personal hygiene and the appropriate use of protective clothing.

The laboratory is a workplace area where potentially dangerous chemicals are in constant use and must therefore remain solely a workplace area. Other related work, such as any necessary paperwork, should be carried out in an adjacent but specified area where chemicals are not in use. Other activities are totally forbidden in the laboratory area, for example eating, drinking, smoking, *etc.*

Hands should be washed before leaving the laboratory area for any reason because of the risk of contamination from substances that may have been handled in the laboratory.

Protective clothing is equally important. A white coat, or equivalent, should be worn in the laboratory at all times and this should not be removed from the laboratory area, except for laundering. This is particularly important when dealing with biologically hazardous substances, in which case the microbiological high-neck-type laboratory coat may be appropriate. When handling biological substances it is also important to ensure that areas of broken skin that may come into contact with such substances should be covered. Gloves and goggles or face masks should be worn and used as appropriate. Mouth pipetting is totally forbidden.

For rapid and potentially dangerous or explosive reactions, a safety shield should be placed between the equipment and the operator, with the whole system placed in a fume cupboard. Fume cupboards should also be used for processes that involve the use or release of potentially harmful and volatile compounds. Fume cupboards are only effective if used properly. This means that the fan must be operational and the front sash window closed to at least the maximum operational position. This is either marked on the side of the cupboard or the information is available from the installers. While working in the cabinet, the window should be as low as possible, leaving only a minimum reasonable operating space between the floor of the cabinet and the bottom of the sash window. No fume cupboard will operate effectively with the sash fully open. The window should be kept in the closed position at all other times. From time to time the air flow through the cabinet should be checked with an air flow meter to ensure that the minimum air flow required is achieved.

21.6   POST-INCIDENT PROCEDURES

Through good practice and safe handling of chemicals it is hoped that untoward incidents will not occur. However, it must be recognized that there will be accidents even with the operation of safe procedures, and the student must be aware of how to deal with these. Procedures may be divided into two categories, those dealing with an affected individual and those concerned directly with the incident.

An affected individual should be removed from the immediate site of the incident. Qualified first aid or medical assistance should be sought urgently if this is relevant. If there is skin contact with corrosive materials the affected areas should be doused with large volumes of cold running water. Similarly if there is eye contact, the area should be thoroughly washed. If in doubt, large quantities of cold running water should be administered until more expert advice is available. For larger affected areas the individual should be placed under an emergency shower which should be available in all chemistry laboratory areas. Clothing should be removed and reservations due to modesty overridden in favour of safety and treatment. In cases of ingestion, forced vomiting may be appropriate or the administration of an antidote, if this is available. Antidotes, if they exist, should be available where chemicals of specific high toxicity are being used. For instance, if a laboratory uses cyanides, the antidote should be available and laboratory staff should be trained in its use.

The most common chemical accident is spillage, either as a consequence of the breakage of the reagent or chemical container, or as the result of a fracture to a vessel in which a chemical reaction is taking place. In either case the area should be evacuated and a quick assessment made as to whether the spillage can be dealt with by staff available without further risk or injury. If this is so, the spillage should be contained and mopped up by the most appropriate procedure. This may be the use of large quantities of water. If the spillage contains a strong acid, then the use of a weak base such as solid sodium carbonate may be more appropriate. Equally if a strong alkali is present, the use of a weak acid, such as dilute acetic acid, may be appropriate. Water would not be appropriate to remove organic solutions that are not miscible with water. In such situations an absorbing reagent such as vermiculite should be used.

If there is any possibility of further exposure of individuals, then the area should be evacuated and sealed off and the assistance of the appropriate professional services sought. If there is release of obnoxious fumes, or fire, these should *only* be dealt with if there is no further risk to individuals. The emission of obnoxious fumes may be dealt with by opening as many of the windows as possible to create ventilation and through draft, and removing the chemicals or reaction from the source that is causing the production of fumes, for example removing the reaction vessel from the heat source, switching off equipment, *etc.*

Small fires may be dealt with by applying the appropriate fire extinguisher, if there is no risk to laboratory personnel. All laboratory personnel should be aware of the location of fire extinguishers, know which type is appropriate for which type of fire, and apply accordingly. Similarly, laboratory personnel should be aware of the location of other safety aids such as first aid cabinets, fire blankets, fire alarm points, and respiratory and other protective clothing and equipment.

21.7 PROTOCOLS AND PROCEDURES

There is an increase in legislation in many countries relating to good laboratory practice and the safe handling of chemicals. This leads to a requirement of compliance with safety documents, the establishment of protocols for carrying out routine procedures, and the listing of chemicals with their particular hazards and toxic properties. Examples of this in the UK have already been mentioned such as the COSHH requirements, and the Department of Health's documents on the safe handling of biological materials. Increasingly, protocols are required for laboratory registration programmes such as accreditation, good working practices, and total quality management. Many professional chemists may find these legislative requirements imposing upon their professional acumen. Nonetheless, they produce a general awareness of safety requirements, and help to establish acceptable minimum standards.

It is also recognized that there is a need to produce Statements of Operational Procedures. These detail the work (analytical) that is being carried out within a laboratory. Each step of a procedure is described, giving reagents and quantities that are required, incubation times, *etc.* In addition these procedures should include any relevant details concerning the hazards or toxicity of any of the reagents involved, and if necessary procedural details in the case of spillage or an accident. Not only do such details help to maintain safety standards, but also they improve analytical competence by ensuring that different operators follow the same protocol.

Chemists involved in research may find it difficult to describe their work in this way, as these procedures are more applicable to routine processes. Nonetheless, documentation such as this will enhance the safety of operational procedures, and the student needs to be aware of their existence, and when to apply them. Knowledge of the existence of such protocols is important because the student will undoubtedly meet them as his or her career develops.

21.8   BIBLIOGRAPHY

'Code of Practice for the Prevention of Infection in Clinical Laboratories and Post Mortem Rooms'. HMSO, London, 1978.

'Safety in Health Service Laboratories. Safe Working and the Prevention of Infection in Clinical Laboratories – Model Rules for Staff and Visitors', HMSO, London, 1991.

S.G. Luxon, ed. 'Hazards in the Chemical Laboratory'. The Royal Society of Chemistry, London, 1992.

P.H. Bach, S.S. Brown, J.A. Haines, R.M. Joyce, B.C. McKusick, and H.G.J. Worth, eds. 'Chemical Safety Matters'. Cambridge University Press, Cambridge, 1992.

Commission of the European Communities, 'Industrial Health and Safety. International Chemical Safety Cards'. Commission of the European Communities, Luxembourg, 1990.

'Health and Safety Data Sheets'. Merck Ltd, Poole, Dorset, 1995.

A Curriculum of Fundamental Toxicology for Chemists

Increasingly, chemists are faced with legislation requiring assessment of hazard and risk associated with the production, use, and disposal of chemicals. In addition, the general public are concerned about the dangers that they hear may result from the widespread use of chemicals. They look to chemists for explanations and assume that chemists understand such matters. When they find that chemists are ignorant of the potential of chemicals to cause harm, their confidence in the profession is lost and chemophobia may result. Thus, it has become essential to introduce toxicology into chemistry courses. In order to facilitate this, the IUPAC Commission on Toxicology and the IUPAC Committee on the Teaching of Chemistry have drafted a curriculum in fundamental toxicology for chemists and this textbook to support it. The curriculum is detailed below.

It will be seen that the titles in this book are echoed in the curriculum. The complete curriculum as shown would provide a thorough basis in toxicology but the 50 hours suggested will be hard to find in any existing chemistry degree structure. We therefore, suggest a minimum curriculum consisting of Sections 1, 2, 3, and 7 (about 30 student contact hours) which should be easier to introduce. This would correspond approximately to the traditional term in a British university or a half-semester in the American system. Alternatively, it could be taught as a 1 week intensive full-time course. In the same way, the full curriculum can be taught intensively over 2 weeks.

Ultimately, any curriculum will be decided by the faculty members who teach it. Our primary aim in producing this book is to provide the building blocks from which curricula can be constructed appropriate to the variety of chemistry courses that exist around the world. Thus, although we have had the curriculum below in mind in planning this book, we regard it simply as a starting point and we hope that it will evolve in practice to produce curricula tailored to the local needs of innovative chemistry departments and their graduates.

THE DETAILED CURRICULUM

Section 1 Introduction to Toxicology
Hazard, toxicity, and fundamental concepts. Toxicity testing and epidemiology as sources of information.
Toxicokinetics
Exposure, adsorption, distribution, storage, metabolism, and excretion
Toxicodynamics
Fundamental mode of action of toxicants in causing harmful effects

Section 2 Application of Toxicity Data
Toxicity data interpretation
Risk assessment and risk management
Monitoring exposure

Section 3 Specialized Aspects of Toxicology
Mutagenicity
Carcinogenicity
Reproductive toxicity
Immunotoxicity

Section 4 Organ Toxicology
Skin toxicology
Respiratory toxicology
Hepatotoxicity
Nephrotoxicity
Neurotoxicity

Section 5 Complex Aspects of Toxicology
Behavioural toxicology
Ecotoxicity

Section 6 Special Toxicants
Radionuclides
Biocides

Section 7 Safe Handling of Chemicals

Summary and Review

Proposed Timetable

Section 1	Introduction to toxicology	12 hours
Section 2	Application of toxicity data	8 hours
Section 3	Specialized aspects of toxicology	8 hours
Section 4	Organ toxicology	10 hours
Section 5	Complex aspects of toxicology	5 hours
Section 6	Special toxicants	4 hours
Section 7	Safe handling of chemicals	2 hours
	Summary and review	1 hour
	Total	50 hours

APPENDIX 2

Glossary of Terms used in Toxicology

This Glossary is an abridged version of a glossary produced for IUPAC in 1993: J.H. Duffus, ed. Glossary for Chemists of Terms Used in Toxicology (IUPAC Recommendations 1993), *Pure Appl. Chem.*, 1993, **65**, 2003–2122.

NOTES FOR THE USER OF THE GLOSSARY

Toxicologists do not always conform, in their current practice, to the rules laid down by IUPAC. For example, IUPAC requires that abbreviations should not be used as symbols for physical quantities, but by long tradition toxicologists have used the abbreviation LD_{50} and other related abbreviations in this way. The glossary accepts prevailing toxicological usage, but it should not to be taken as an indication that it has IUPAC approval. In contrast, and according to IUPAC practice, but contrary to much current usage, toxic components (like other chemically definable components) in doses, dosages, and exposures are expressed as amounts of substance, substance concentration, substance content, or a time integral of one of these. Thus, for example, a pesticide should be expressed as amount of substance, not as its mass including binding agents and other ingredients, nor its volume including solvent.

Throughout the glossary the following abbreviations are used to indicate the relationships between terms.

AN antonym, opposite
BT broader term
NT narrower term
PS partial synonym
RT related term
SN exact synonym

Abiotic transformation   Process in which a substance in the environment is modified by non-biological mechanisms. RT **biotransformation**.

Absolute lethal concentration (LC$_{100}$)   Lowest concentration of a substance in an environmental medium that kills 100% of test organisms or species under defined conditions. This value is dependent on the number of organisms used in its assessment.

Absolute lethal dose (LD$_{100}$)   Lowest amount of a substance that kills 100% of test animals under defined conditions. This value is dependent on the number of organisms used in its assessment.

Absorbed dose (of a substance)   Amount of a substance absorbed into an organism or into organs and tissues of interest.

Absorbed dose (of radiation)   Energy imparted to matter in a suitably small element of volume by ionizing radiation divided by the mass of that element of volume. The SI unit for absorbed dose is joule per kilogram (J kg^{-1}) and its special name is gray (Gy). RT **ionizing radiation**.

Absorption (biological)   Process of active or passive transport of a substance into an organism: in the case of a mammal or human being, this is usually through the lungs, gastrointestinal tract, or skin.

Absorption (of radiation)   Phenomenon in which radiation transfers some or all of its energy to matter that it traverses.

Acaricide   Substance intended to kill mites, ticks, or other Acaridae.

Acceptable daily intake (ADI)   Estimate of the amount of a substance in food or drinking water, expressed on a body mass basis (usually mg kg^{-1} body weight), that can be ingested daily over a lifetime by humans without appreciable health risk. For calculation of the daily intake per person, a standard body mass of 60 kg is used. ADI is normally used for food additives (tolerable daily intake is used for contaminants). RT **tolerable daily intake**.

Acceptable risk   Probability of suffering disease or injury which is considered to be sufficiently small to be 'negligible'. PS **tolerable risk**. RT **accepted risk, negligible risk, risk *de minimis*.**

Accumulation   Successive additions of a substance to a target organism, or organ, or to part of the environment, resulting in an increasing amount or concentration of the substance in the organism, organ, or environment.

Acidosis   Pathological condition in which the hydrogen ion substance concentration of body fluids is above normal and the pH of blood falls below the reference interval. AN **alkalosis**.

Activation   See NT **bioactivation**

Acute
1. Short-term, in relation to exposure or effect. In experimental toxicology, 'acute' refers to studies of 2 weeks or less in duration (often less than 24 h). AN **chronic**
2. In clinical medicine, sudden and severe, having a rapid onset.

Acute toxicity
1. Adverse effects occurring within a short time (usually up to 14 days) after administration of a single dose (or exposure to a given concentration) of a test substance, or after multiple doses (exposures), usually within 24 h.
2. Ability of a substance to cause adverse effects within a short time of dosing or exposure. AN **chronic toxicity**.

Acute toxicity test   Experimental animal study to determine what adverse effects occur in a short time (usually up to 14 days) after a single dose of a substance or after multiple doses given in up to 24 h. RT **limit test, median lethal dose (LD$_{50}$)**.

Addiction   Surrender and devotion to the regular use of a medicinal or pleasurable substance for the sake of relief, comfort, stimulation, or exhilaration which it affords; often with craving when the drug is absent. PS **dependence**.

Additive effect   Consequence that follows exposure to two or more physicochemical agents which act jointly but do not interact: commonly, the total effect is the simple sum of the effects of separate exposure to the agents under the same conditions. Substances of similar simple action may show dose or concentration addition. RT **antagonism, combined effect of poisons, potentiation, synergism**.

Adenocarcinoma   Malignant tumour originating in glandular epithelium or forming recognizable glandular structures. RT **adenoma**.

Adenoma   Benign tumour occurring in glandular epithelium or forming recognizable glandular structures. RT **adenocarcinoma**.

Adjuvant
1. In pharmacology, a substance added to a drug to speed or increase the action of the main component.
2. In immunology, a substance (such as aluminium hydroxide) or an organism (such as bovine tuberculosis bacillus) which increases the response to an antigen.

Adrenergic   See SN **sympathomimetic**

Adverse effect Change in morphology, physiology, growth, development, or life span of an organism which results in impairment of functional capacity or impairment of capacity to compensate for additional stress, or increase in susceptibility to the harmful effects of other environmental influences.

Aerobe Organism that needs molecular oxygen for respiration and hence for growth and life. AN **anaerobe**.

Aerodynamic diameter (of a particle) Diameter of a spherical particle of unit density which has the same settling velocity in air as the particle in question.

Aerosol Dispersion of liquid or solid material in a gas.

Aetiology
1. Science dealing with the cause or origin of disease.
2. In individuals, the cause or origin of disease. RT **epidemiology**.

Agonist Substance that binds to cell receptors normally responding to naturally occurring substances and which produces a response of its own. AN **antagonist**.

Albuminuria Presence of albumin, derived from plasma, in the urine. RT **microalbuminuria, proteinuria**.

Algicide Substance intended to kill algae.

Alkalosis Pathological condition in which the hydrogen ion substance concentration of body fluids is below normal and the pH of blood rises above the reference interval. AN **acidosis**.

Alkylating agent Substance that introduces an alkyl substituent into a compound.

Allele One of several alternate forms of a gene which occurs at the same relative position (locus) on homologous chromosomes and which becomes separated during meiosis and can be recombined following fusion of gametes. RT **gametes, meiosis**.

Allergen Antigenic substance capable of producing immediate hypersensitivity. RT **allergy, antigen, hypersensitivity**.

Allergy Symptoms or signs occurring in sensitized individuals following exposure to a previously encountered substance (allergen) which would otherwise not cause such symptoms or signs in non-sensitized individuals. The most common forms of allergy are rhinitis, urticaria, asthma, and contact dermatitis. RT **immune response, hypersensitivity**.

All-or-none effect   See SN **quantal effect**. RT **stochastic effect**

Alveol/us (pulmonary), **-i** pl., **-ar** adj. Terminal air sac of the lung where gas exchange occurs.

Ambient monitoring   Continuous or repeated measurement of agents in the environment to evaluate ambient exposure and health risk by comparison with appropriate reference values based on knowledge of the probable relationship between exposure and resultant adverse health effects. RT **biological monitoring**, **environmental monitoring**, **monitoring**.

Ambient standard   See SN **environmental quality standard**

Ames test *In vitro* test for mutagenicity using mutant strains of the bacterium *Salmonella typhimurium* which cannot grow in a given histidine-deficient medium: mutagens can cause reverse mutations which enable the bacterium to grow on the medium. The test can be carried out in the presence of a given microsomal fraction (S-9) from rat liver to allow metabolic transformation of mutagen precursors to active derivatives.

Amplification (of genes)   See **gene amplification**

Anabolism   Biochemical processes by which smaller molecules are joined to make larger molecules. AN **catabolism**.

Anaerobe   Organism that does not need molecular oxygen for life. Obligate (strict) anaerobes grow only in the absence of oxygen. Facultative anaerobes can grow either in the presence or in the absence of molecular oxygen. AN **aerobe**.

Anaesthetic   Substance that produces loss of feeling or sensation: general anaesthetic produces loss of consciousness; local or regional anaesthetic renders a specific area insensible to pain.

Analgesic   Substance that relieves pain, without causing loss of consciousness.

Analytic study (in epidemiology)   Hypothesis-testing method of investigating the association between a given disease or health state or other dependent variable and possible causative factors. In an analytic study, individuals in the study population are classified according to absence or presence (or future development) of specific disease and according to attributes that may influence disease occurrence. Attributes may include age, race, sex, other disease(s), genetic, biochemical, and physiological characteristics, economic status, occupation, residence, and various aspects of the environment or personal behaviour. Three types of analytic study are: cross-sectional (prevalence), cohort (prospective), and case control (retrospective).

Anaphylaxis Severe allergic reaction occurring in a person or animal exposed to an antigen or hapten to which they have previously been sensitized. RT **antigen, hapten**.

Anaplasia Loss of normal cell differentiation; this is a feature characteristic of most malignancies. RT **malignancy**.

Anemia See **anaemia**

Aneuploid Cell or organism with missing or extra chromosomes or parts of chromosomes.

Anoxia Strictly, total absence of oxygen, but sometimes used to mean decreased oxygen supply in tissues.

Antagonism Combined effect of two or more factors which is smaller than the solitary effect of any one of those factors. In bioassays, the term is used when a specified response is produced by exposure to either of two factors but not by exposure to both together. RT **synergism**.

Antagonist
1. Substance that reverses or reduces the effect induced by an agonist.
2. Substance that attaches to and blocks cell receptors that normally bind naturally occurring substances. AN **agonist**.

Anthelmint(h)ic Substance intended to kill parasitic intestinal worms, such as helminths.

Anti-adrenergic See SN **sympatholytic**

Antibiotic Substance produced by, and obtained from, certain living cells (especially bacteria, yeasts, and moulds), or an equivalent synthetic substance, which is biostatic or biocidal at low concentrations to some other form of life, especially pathogenic or noxious organisms.

Antibody Protein molecule produced by the cells of the immune system (an immunoglobulin molecule) that can bind specifically to the molecule (antigen or hapten) which induced its synthesis. RT **antigen, hapten, immunoglobulin**.

Anticholinergic
1. adj. Preventing transmission of parasympathetic nerve impulses.
2. n. Substance that prevents transmission of parasympathetic nerve impulses.

Anticholinesterase See SN **cholinesterase inhibitor**.

Anticoagulant   Substance that prevents clotting.

Antidote   Substance capable of specifically counteracting or reducing the effect of a potentially toxic substance in an organism by a relatively specific chemical or pharmacological action.

Antigen   Substance or a structural part of a substance that causes the immune system to produce specific antibody or specific cells and which combines with specific binding sites (epitopes) on the antibody or cells. RT **antibody**, **epitope**.

Antihelminth   See SN **anthelmint(h)ic**

Antimetabolite   Substance, structurally similar to a metabolite, that competes with it or replaces it, and so prevents or reduces its normal utilization.

Antimycotic   Substance used to kill a fungus or to inhibit its growth. SN **fungicide**.

Antipyretic   Substance that relieves or reduces fever.

Anti-resistant   Substance used as an additive to a pesticide formulation in order to reduce the resistance of insects to the pesticide.

Antiserum   Serum containing antibodies to a particular antigen either because of immunization or after an infectious disease.

Aphasia   Loss or impairment of the power of speech or writing, or of the ability to understand written or spoken language or signs, due to a brain injury or disease.

Aphicide   Substance intended to kill aphids. BT **insecticide**.

Aplasia   Lack of development of an organ or tissue, or of the cellular products from an organ or tissue.

Apoptosis   Physiological process of programmed tissue death (and disintegration) associated with normal development in animals. RT **necrosis**.

Arboricide   Substance intended to kill trees and shrubs.

Arrhythmia   Any variation from the normal rhythm of the heartbeat.

Artefact   Findings or product of experimental or observational techniques that is not properly associated with the system being studied.

Arteriosclerosis Hardening and thickening of the walls of the arteries.

Arthritis Inflammation of a joint, usually accompanied by pain and often by changes in structure.

Asbestosis Type of pneumoconiosis caused by inhaled asbestos fibres. BT **pneumoconiosis**.

Ascaricide Substance intended to kill roundworms (Ascaridae).

Asphyxia Condition resulting from insufficient intake of oxygen: symptoms include breathing difficulty, impairment of senses, and, in extreme cases, convulsions, unconsciousness, and death.

Asphyxiant Substance that blocks the transport or use of oxygen by living organisms.

Assessment Factor See SN **uncertainty factor**.

Asthenia Weakness; lack or loss of strength.

Asthma Chronic respiratory disease characterized by bronchoconstriction, excessive mucus secretion, and oedema of the pulmonary alveoli, resulting in difficulty in breathing out, wheezing, and cough.

Astringent
1. adj. Causing contraction, usually locally after topical application.
2. n. Substance causing cells to shrink, thus causing tissue contraction or stoppage of secretions and discharges; such substances may be applied to skin to harden and protect it.

Ataxia Unsteady or irregular manner of walking or movement caused by loss or failure of muscular co-ordination.

Atherosclerosis Pathological condition in which there is thickening, hardening, and loss of elasticity of the walls of blood vessels, characterized by a variable combination of changes of the innermost layer consisting of local accumulation of lipids, complex carbohydrates, blood and blood components, fibrous tissue, and calcium deposits. In addition, the outer layer becomes thickened and there is fatty degeneration of the middle layer.

Atrophy Wasting away of the body or of an organ or tissue.

Autoimmune disease Pathological condition resulting when an organism produces antibodies or specific cells that bind to constituents of its own tissues (autoantigens) and cause tissue injury: examples of such disease may include rheumatoid arthritis, myasthenia gravis, and scleroderma. RT **allergy**, **antibody**, **antigen**, **hypersensitivity**, **immune response**.

Autophagosome   Membrane-bound body (secondary lysosome) in which parts of the cell are digested.

Autopsy   Post-mortem examination of the organs and body tissue to determine cause of death or pathological condition. RT **biopsy**. SN **necropsy**.

Avicide   Substance intended to kill birds.

Axenic animal   See SN **germ-free animal**

Back-mutation   Process that reverses the effect of a mutation that had inactivated a gene; thus it restores the wild phenotype. RT **phenotype**.

Bactericide   Substance intended to kill bacteria.

B-cell   See **B-lymphocyte**

Benign
1. Of a disease, producing no persisting harmful effects.
2. Tumour that does not invade other tissues (metastasize), having lost growth control but not positional control. AN **malignant**.

Berylliosis   See SN **beryllium disease**

Beryllium disease   Serious and usually permanent lung damage resulting from chronic inhalation of beryllium.

Bilirubin   Orange-yellow pigment ($C_{33}H_{36}O_6N_4$), a breakdown product of haeme-containing proteins (haemoglobin, myoglobin, cytochromes), which circulates in the blood plasma bound to albumin or as water-soluble glucuronides, and is excreted in the bile by the liver.

Bioaccumulation   Progressive increase in the amount of a substance in an organism or part of an organism which occurs because the rate of intake exceeds the organism's ability to remove the substance from the body. PS **bioconcentration**, **biomagnification**.

Bioaccumulation potential   Ability of living organisms to concentrate a substance obtained either directly from the environment or indirectly through its food.

Bioactivation   Metabolic conversion of a xenobiotic to a more toxic derivative. PS **activation**. BT **biotransformation**.

Bioassay Procedure for estimating the concentration or biological activity of a substance (vitamin, hormone, plant growth factor, antibiotic, *etc.*) by measuring its effect on an organism compared with an appropriate standard preparation. BT **assay**.

Bioavailability
1. Extent to which a substance that the body is exposed to (by ingestion, inhalation, injection, or skin contact) reaches the systemic circulation, and the rate at which this occurs. SN **biological availability, physiological availability**.
2. Pharmacokinetic term relating systemic exposure from extravascular exposure (ev) to that following intravenous exposure (iv) by the equation:

$$F = \text{AUC}_{ev} \; D_{iv}/\text{AUC}_{iv} \; D_{ev}$$

where F is the bioavailability, AUC_{ev} and AUC_{iv} are the areas under the plasma concentration–time curve following extravascular and intravenous administration, and D_{ev} and D_{iv} are the administered extravascular and intravenous doses.

Biochemical (biological) oxygen demand (BOD) Substance concentration of oxygen taken up through the respiratory activity of micro-organisms growing on organic compounds present when incubated at a specified temperature (usually 20 °C) for a fixed period (usually 5 days). It is regarded as a measure of that organic pollution of water which can be degraded biologically but includes the oxidation of inorganic material such as sulfide and iron(II). The empirical test used in the laboratory to determine BOD also measures the oxygen used to oxidize reduced forms of nitrogen unless their oxidation is prevented by an inhibitor such as allyl thiourea. RT **chemical oxygen demand**.

Biocid/e n., **-al** adj. Substance intended to kill living organisms.

Bioconcentration Process leading to a higher concentration of a substance in an organism than in environmental media to which it is exposed. PS **bioaccumulation, biomagnification**.

Bioconcentration factor (BCF) Measure of the tendency for a substance in water to accumulate in fish tissue or in tissues of other organisms. The equilibrium concentration of a substance in fish can be estimated by multiplying the concentration of the substance in the surrounding water by the fish bioconcentration factor for that chemical. This parameter is an important determinant for human intake by the aquatic food ingestion route.

Biodegradation Breakdown of a substance catalysed by enzymes *in vitro* or *in vivo*. This may be characterized for purposes of hazard assessment as:
1. Primary. Alteration of the chemical structure of a substance resulting in loss of a specific property of that substance.

2. Environmentally acceptable. Biodegradation to such an extent as to remove undesirable properties of the compound. This often corresponds to primary biodegradation but it depends on the circumstances under which the products are discharged into the environment.
3. Ultimate. Complete breakdown of a compound to either fully oxidized or reduced simple molecules (such as carbon dioxide/methane, nitrate/ammonium, and water).

It should be noted that the products of biodegradation can be more harmful than the substance degraded. RT **biotransformation**.

Biological assessment of exposure   Assessment of exposure to a substance by the analysis of specimens taken in the environment such as foodstuffs, plants, animals, biological material in air or water samples, or biological material from exposed subjects. When human samples are analysed, they are usually urine and blood; other possible samples include expired air, faeces, saliva, bile, hair, and biopsy or autopsy material. When other organisms are being considered, the whole organism may be analysed as well as selected tissues such as fat in pigs or birds. In these samples, the content(s) of the substance(s) or metabolite(s) is determined and, on this basis, the exposure level (concentration in the air, absorbed amount of the substance) or the probability of health impairment due to exposure are derived. Biochemical changes in the components of an organism, such as changes in enzyme activity or in the excretion of metabolic intermediates, can also be used for this purpose if they show a relationship to the exposure. BT **biological monitoring, monitoring**.

Biological effect monitoring (BEM)   Continuous or repeated measurement of early biological effects of exposure to a substance to evaluate ambient exposure and health risk by comparison with appropriate reference values based on knowledge of the probable relationship between ambient exposure and biological effects. BT **biological monitoring, environmental monitoring**.

Biological half-life or half-time $(t_{1/2})$ Time required for the amount of a substance in a biological system to be reduced to one-half, predominantly by biological processes, when the rate of removal is approximately exponential.

Biological monitoring   Continuous or repeated measurement of potentially toxic substances or their metabolites or biochemical effects in tissues, secreta, excreta, expired air, or any combination of these in order to evaluate occupational or environmental exposure and health risk by comparison with appropriate reference values based on knowledge of the probable relationship between ambient exposure and resultant adverse health effects. NT **biological effect monitoring**. BT **environmental monitoring, monitoring**. RT **biological assessment of exposure**.

Biological oxygen demand   See SN **biochemical oxygen demand**

Biomagnification Sequence of processes in an ecosystem by which higher concentrations are attained in organisms at higher trophic levels (at higher levels in the food web); at its simplest, a process leading to a higher concentration of a substance in an organism than in its food. SN **ecological magnification**. RT **bioaccumulation, bioconcentration**.

Biomarker
1. Parameter that can be used to identify a toxic effect in an individual organism and can be used in extrapolation between species.
2. Indicator signalling an event or condition in a biological system or sample and giving a measure of exposure, effect, or susceptibility.

Biomass
1. Total amount of biotic material, usually expressed per unit surface area or volume, in a medium such as water.
2. Material produced by the growth of micro-organisms, plants, or animals.

Biomineralization Complete conversion of organic substances to inorganic derivatives by living organisms, especially micro-organisms.

Biomonitoring See SN **biological monitoring**

Biopsy Excision of a small piece of living tissue for microscopic or biochemical examination; usually performed to establish a diagnosis. RT **autopsy**.

Biosphere Portion of the planet earth that supports and includes life.

Biota All living organisms as a totality.

Biotransformation Any chemical conversion of substances that is mediated by living organisms or enzyme preparations derived therefrom.

B-lymphocyte Type of lymphocyte that synthesizes and secretes antibodies in response to the presence of a foreign substance or one identified by it as foreign. The protective effect can be mediated to a certain extent by the antibody alone (contrast T lymphocyte). RT **immune response, lymphocyte, T-lymphocyte**.

Body burden Total amount of substance of a chemical present in an organism at a given time.

Bolus
1. Single dose of a substance, originally a large pill.
2. Dose of a substance administered by a single, rapid, intravenous injection.
3. Concentrated mass of food ready to be swallowed.

Brady-   Prefix meaning slow as in bradycardia or bradypnoea.

Bradycardia   Abnormal slowness of the heartbeat. AN **tachycardia**.

Bradypnoea   Abnormally slow breathing. AN **tachypnoea**.

Breathing zone   Space within a radius of 0.5 m from a person's face.

Bronchoconstriction   Narrowing of the air passages through the bronchi of the lungs. AN **bronchodilation**.

Bronchodilation   Expansion of the air passages through the bronchi of the lungs. AN **bronchoconstriction**.

Bronchospasm   Intermittent violent contraction of the air passages of the lungs.

Cancer   Disease resulting from the development of a malignancy. RT **carcinogen**, **carcinogenesis**, **carcinogenic**, **carcinogenicity**, **carcinoma**, **malignant**, **malignancy**.

Carboxyhaemoglobin   Compound that is formed between carbon monoxide and haemoglobin in the blood of animals and is incapable of transporting oxygen.

Carcinogen n., **-ic** adj. Agent (chemical, physical, or biological) that is capable of increasing the incidence of malignant neoplasms; the induction of benign neoplasms may in some circumstances contribute to the judgement that an agent is carcinogenic.

Carcinogen/esis n., **-etic** adj. Induction, by chemical, physical, or biological agents, of malignant neoplasms.

Carcinogenicity   Process of induction of malignant neoplasms by chemical, physical, or biological agents.

Carcinogenicity test   Long-term (chronic) test designed to detect any possible carcinogenic effect of a test substance.

Carcinoma   Malignant tumour of an epithelial cell. BT **epithelioma**.

Cardiotoxic   Chemically harmful to the cells of the heart.

Case control study   A study that starts with the identification of persons with the disease (or other outcome variable) of interest, and a suitable control (comparison, reference) group of persons without the disease. The relationship of an attribute to the disease is examined by comparing the diseased and non-diseased with regard to how frequently the attribute is present or, if quantitative, the levels of the attribute, in the two groups. SN **case comparison study, case history study, case referent study, retrospective study**.

Ceiling value (CV)   US term in occupational exposure indicating the airborne concentration of a potentially toxic substance which should never be exceeded in a worker's breathing zone.

Cell line   Defined unique population of cells obtained by culture from a primary implant through numerous generations.

Cell-mediated hypersensitivity   State in which an individual reacts with allergic effects caused by the reaction of antigen-specific T-lymphocytes following exposure to a certain substance (allergen) after having been exposed previously to the same substance or chemical group. RT **allergy, antigen, immunoglobulin E-mediated hypersensitivity**.

Cell-mediated immunity   Immune response mediated by antigen-specific T-lymphocytes.

Cell strain   Cells having specific properties or markers derived from a primary culture or cell line.

Chemical oxygen demand (COD)   Substance concentration of available oxygen (derived from a chemical oxidizing agent) required to oxidize the organic (and inorganic) matter in waste water. RT **biochemical oxygen demand**.

Chemical species   Set of chemically identical atomic or molecular structural units in a solid array or of chemically identical molecular entities that can explore the same set of molecular energy levels on the time scale of the experiment. For example, two conformational isomers may interconvert sufficiently slowly to be detectable by separate NMR spectra and therefore be considered to be separate chemical species on a time scale governed by the radiofrequency of the spectrometer used. However, in a slow chemical reaction the same mixture of conformers may behave as a single chemical species, *i.e.* there is a virtually complete equilibrium population of the total set of molecular energy levels belonging to the two conformers. Except where the context requires otherwise, the term is taken to refer to a set of molecular entities containing isotopes in their natural abundance. The wording of the definition given is intended to embrace both cases such as graphite, sodium chloride, or a surface oxide where the basic structural units are not capable of a separate existence as well as those cases where they are.

Chemobiokinetics   See NT **toxicokinetics**
Chemophobia   Irrational fear of chemicals.

Chloracne Acne-like eruption caused by exposure to certain chlorinated organic substances such as polychlorinated biphenyls or 2,3,7,8–tetrachlorodibenzo-*p*-dioxin.

Cholinesterase inhibitor Substance that inhibits the action of acetylcholinesterase (EC 3.1.1.7) and related enzymes which catalyse the hydrolysis of choline esters: such a substance causes hyperactivity in parasympathetic nerves.

Cholinomimetic See SN **parasympathomimetic**

Chromatid Either of two filaments joined at the centromere that make up a chromosome.

Chromatin Stainable complex of DNA and proteins present in the nucleus of a eukaryotic cell. RT **eukaryote**.

Chromosomal aberration Abnormality of chromosome number or structure.

Chromosome Self-replicating structure consisting of DNA complexed with various proteins and involved in the storage and transmission of genetic information; the physical structure that contains the genes. RT **chromatid**.

Chronic exposure Continued exposures occurring over an extended period of time, or a significant fraction of the test species' or of the group of individuals', or of the population's lifetime. AN **acute exposure**.

Chronic toxicity
1. Adverse effects following chronic exposure.
2. Effects that persist over a long period of time whether or not they occur immediately upon exposure or are delayed. AN **acute toxicity**.

Chronic toxicity test Study in which organisms are observed during the greater part of the life span and in which exposure to the test agent takes place over the whole observation time or a substantial part thereof. AN **acute toxicity test**. SN **long-term test**.

Chronotoxicology Study of the influence of biological rhythms on the toxicity of substances.

Cirrhosis
1. Liver disease defined by histological examination and characterized by increased fibrous tissue, abnormal physiological changes such as loss of functional liver cells, and increased resistance to blood flow through the liver (portal hypertension).

2. Interstitial fibrosis of an organ.

Clastogen Agent causing chromosome breakage and/or consequent gain, loss, or rearrangement of pieces of chromosomes.

Clastogenesis Occurrence of chromosomal breaks and/or consequent gain, loss, or rearrangement of pieces of chromosomes.

Clearance
1. Volume of blood or plasma or mass of an organ effectively cleared of a substance by elimination (metabolism and excretion) in a given time interval: clearance is expressed in units of volume or mass per unit of time. Total clearance for a component is the sum of the clearances of each eliminating organ or tissue for that component.
2. In pulmonary toxicology, clearance refers specifically to removal of any inhaled substance which deposits on the lining surface of the lung: lung clearance is expressed in volume or mass of lung cleared per unit time.
3. In renal toxicology, clearance refers to the quantification of the removal of a substance by the kidneys by the processes of filtration and secretion: clearance is calculated by relating the rate of renal excretion to the plasma concentration.
RT **elimination**.

Clon/e nv., **-al** adj.
1. Population of genetically identical cells or organisms having a common ancestor.
2. To produce such a population.
3. Recombinant DNA molecules all carrying the same inserted sequence.

Clonic Pertaining to alternate muscular contraction and relaxation in rapid succession. RT **tonic**.

Cocarcinogen Chemical, physical, or biological factor that intensifies the effect of a carcinogen.

Cohort Component of the population born during a particular period and identified by period of birth so that its characteristics (such as causes of death and numbers still living) can be ascertained as it enters successive time and age periods. The term 'cohort' has broadened to describe any designated group of persons followed or traced over a period of time, as in the term cohort study (prospective study).

Cohort analysis Tabulation and analysis of morbidity or mortality rates in relationship to the ages of a specific group of people (cohort), identified by their birth period, and followed as they pass through different ages during part or all of their life span. In certain circumstances such as studies of migrant populations, cohort analysis may be performed according to duration of residence in a country rather than year of birth, in order to relate health or mortality experience to duration of exposure.

Cohort study   Method of epidemiological study in which subsets of a defined population can be identified who are, have been, or in the future may be exposed or not exposed, or exposed in different degrees, to a factor or factors hypothesized to influence the probability of occurrence of a given disease or other outcome. Alternative terms for such a study—follow-up, longitudinal, and prospective study—describe an essential feature of the method, observation of the population for a sufficient number of person-years to generate reliable incidence or mortality rates in the population subsets. This generally means studying a large population, study for a prolonged period (years), or both. SN **concurrent study, follow-up study, incidence study, longitudinal study, prospective study.**

Co-metabolism   Process by which a normally non-biodegradable substance is biodegraded only in the presence of an additional carbon source. RT **analogue metabolism**.

Compartment   Part of the body considered as an independent system for purposes of assessment of distribution and clearance of a substance. The body is composed of a large number of organs, tissues, cells, cell organelles and fluids, any one of which could be referred to as a compartment. In kinetic considerations, a compartment often refers collectively to the organs, tissues, cells, and fluids for which the rates of uptake and subsequent distribution and elimination are sufficiently similar to preclude kinetic resolution.

Compensation   Adaptation of an organism to changing conditions of the environment (especially chemical) is accompanied by the emergence of stresses in biochemical systems that exceed the limits of normal (homeostatic) mechanisms. Compensation is a temporary concealed pathology that later on can be manifested in the form of explicit pathological changes (decompensation). SN **pseudoadaptation**. RT **acclimatization, adaptation**.

Complete mineralization   Complete breakdown of a complex organic compound to carbon dioxide, water, oxides, and oxidative inorganic products such as nitrate or sulfate.

Concentration (amount-of-substance concentration) $c = n/V$ Derived kind-of-quantity defined as the amount of substance (n) of a component specified by an elementary entity divided by the volume (V) of the system containing the component. The fundamental unit is mol m^{-3} but practical units are mol dm^{-3} or mol L^{-1} (*not molarity*). RT **absolute lethal concentration, lethal concentration, maximum tolerable concentration, median effective concentration, median lethal concentration, median narcotic**

concentration, minimum lethal concentration, threshold concentration.

Concentration–effect curve   Graph of the relation between exposure concentration and the magnitude of resultant biological change. RT **dose–effect curve**. SN **exposure–effect curve**.

Concentration–effect relationship   Association between exposure concentration and the magnitude of the resultant continuously graded change, either in an individual or in a population. RT **dose–effect relationship**.

Concentration–response curve   Graph of the relation between exposure concentration and the proportion of individuals in a population responding with a quantal effect. RT **dose–response curve, response**.

Concentration–response relationship   Association between exposure concentration and the incidence of a defined biological effect in an exposed population. RT **dose–response relationship, response**.

Confounding
  1. Situation in which the effects of two processes are not distinguishable from one another: the distortion of the apparent effect of an exposure on risk brought about by the association of other factors that can influence the outcome.
  2. Relationship between the effects of two or more causal factors, as observed in a set of data, such that it is not logically possible to separate the contribution which any single causal factor has made to an effect.
  3. Situation in which a measure of the effect of an exposure on risk is distorted because of the association of exposure with other factor(s) that influence the outcome under study.

Confounding variable   Changing factor that can cause or prevent the outcome of interest, is not an intermediate variable, and is not associated with the factor under investigation: such a variable must be controlled in order to obtain an undistorted estimate of the effect of the study factor on risk. SN **confounder**.

Congener   Substance that by structure, function, or origin is similar to another.

Conjunctiva   Mucous membrane that covers the eyeball and lines the undersurface of the eyelid.

Conjunctivitis   Inflammation of the conjunctiva.

Conservative assessment of risk   Assessment of risk that assumes the worst possible case scenario and therefore gives the highest possible value for risk: risk management decisions based on this value will maximize safety.

Contact dermatitis   Inflammatory condition of the skin resulting from dermal exposure to an allergen (sensitizer) or an irritating (corrosive, defatting) substance.

Contra-indication   Any condition that renders some particular line of treatment improper or undesirable.

Control group   Selected group, identified as a rule before a study is done, that comprises humans, animals, or other species who do not have the disease, intervention, procedure, or whatever is being studied, but in all other respects is as nearly identical to the test group as possible. SN **comparison group**.

Control, matched   Control (individual or group or case) selected to be similar to a study individual or group or case, in specific characteristics: some commonly used matching variables are age, sex, race, and socio-economic status.

Corrosive   Causing a surface-destructive effect on contact; in toxicology, this normally means causing visible destruction of the skin, eyes, or the lining of the respiratory tract or the gastrointestinal tract.

Count mean diameter   Mean of the diameters of all particles in a population.

Count median diameter   Calculated diameter in a population of particles in a gas or liquid phase above which there are as many particles with larger diameters as there are particles below it with smaller diameters.

Crepitations   Abnormal respiratory sounds heard on auscultation of the chest, produced by passage of air through passages that contain secretion or exudate or that are constricted by spasm or a thickening of their walls (Auscultation is the process of listening for sounds within the body by ear unassisted or using a stethoscope.) SN **crackles**, **râles**.

Critical concentration (for a cell or organ)   Concentration of a potentially toxic substance at which undesirable (or adverse) functional changes, reversible or irreversible, occur in the cell or organ.

Critical effect   For deterministic effects, the first adverse effect that appears when the threshold (critical) concentration or dose is reached in the critical organ. Adverse effects, such as cancer, with no defined threshold concentration are often regarded as critical. Decision on whether an effect is critical is a matter of expert judgement.

Critical end-point   Toxic effect used by the US Environmental Protection Agency as the basis for a reference dose. RT **reference dose**.

Critical group   Part of a target population most in need of protection because it is most susceptible to a given toxicant.

Critical organ
   1. In toxicology. Organ that first attains the critical concentration (of a potentially toxic substance) under specified circumstances of exposure and for a given population.
   2. In radiation biology. Organ, the damage of which (by radiation) results in the greatest injury to the individual (or his/her descendants). The injury may result from inherent radiosensitivity or indispensability of the organ, or from high dose, or from a combination of all three.

Critical organ concentration (of a substance)   Mean concentration in the critical organ at the time the most sensitive type of cell reaches the critical concentration. RT **critical concentration, critical organ**.

Critical period (of development)   Stage of development of an organism that is of particular importance in the life cycle if the normal full development of some anatomical, physiological, metabolic, or psychological structure or function is to be attained: such a period may be associated with very high susceptibility to specific potentially toxic substances.

Cross-sectional study (of disease prevalence and associations)   Study that examines the relationship between diseases (or other health-related characteristics) and other variables of interest as they exist in a defined population at one particular time. Disease prevalence rather than incidence is normally recorded in a cross-sectional study and the temporal sequence of cause and effect cannot necessarily be determined. SN **disease frequency survey, prevalence study**. RT **morbidity survey**.

Cumulative effect   Overall adverse change that occurs when repeated doses of a harmful substance or radiation have biological consequences that are mutually enhancing. SN **functional accumulation**.

Cumulative incidence, cumulative incidence rate   Number and proportion of a group of people who experience the onset of a health-related event during a specified time interval; this interval is generally for all members of the group, but, as in lifetime incidence, it may vary from person to person without reference to age.

Cumulative incidence ratio   Value obtained by dividing the cumulative incidence rate in the exposed population by the cumulative incidence rate in the unexposed population.

Cumulative median lethal dose   Estimate of the total administered amount of a substance that is associated with the death of half a population of animals when the substance is administered repeatedly in doses which are generally fractions of the median lethal dose. The estimate may vary with the chosen size of the fraction (0.1, 0.2 *etc.*) and with the period of time over which effects are observed. It is a calculated quantity generally obtained by interpolation of available dose–response data relating the total administered amount to the response in the corresponding group of experimental animals. BT **median lethal dose**

Cutaneous   Pertaining to the skin. SN **dermal**.

Cyanogenic   Compounds able to produce cyanide; examples are the cyanogenic glycosides such as amygdalin in peach and apricot stones.

Cyanosis   Bluish coloration, especially of the skin, mucous membranes, and fingernail beds, caused by abnormally large amounts of reduced haemoglobin in the blood vessels as a result of deficient oxygenation.

Cytochrome   Haemoprotein whose characteristic mode of action involves transfer of reducing equivalents associated with a reversible change in oxidation state of the haeme prosthetic group: strictly, the cytochrome P450 family are not cytochromes but haeme–thiolate proteins.

Cytochrome P420   Inactive derivative of cytochrome P450 found in microsomal preparations. RT **cytochrome P448, cytochrome P450, endoplasmic reticulum, microsome, monooxygenase, phase 1 reaction**.

Cytochrome P448   Obsolete term for cytochrome P450 I, A1, and A2, one of the major families of the cytochrome P450 haemoproteins. During the monooxygenation of certain substances, often a detoxification process, these isoenzymes may produce intermediates that initiate mutations, chemical carcinogenesis, immunotoxic reactions, and other forms of chemical toxicity. RT **cytochrome P420, cytochrome P450, endoplasmic reticulum, microsome, monooxygenase, phase 1 reaction**.

Cytochrome P450   Haemoproteins that form the major part of the enzymes concerned with the monooxygenation of many endogenous and exogenous substrates. The term includes a large number of isoenzymes that are coded for by a superfamily of genes. Endogenous substrates of these enzymes include cholesterol, steroid hormones, and the eicosenoids; the exogenous substrates are xenobiotics. Strictly, the cytochrome P450 family are not cytochromes but are haeme–thiolate proteins. SN **mixed-function oxidase**. RT **cytochrome P420, cytochrome P448, endoplasmic reticulum, microsome, monooxygenase, phase 1 reaction, xenobiotic**.

Cytogenetics Branch of genetics that correlates the structure and number of chromosomes as seen in isolated cells with variation in genotype and phenotype. RT **phenotype**.

Cytotoxic Causing damage to cell structure or function.

Defoliant Substance used for removal of leaves by its toxic action on living plants.

Denaturation
1. Addition of methanol or acetone to alcohol to make it unfit for drinking.
2. Change in molecular structure of proteins so that they cannot function normally, often caused by splitting of hydrogen bonds following exposure to reactive substances or heat.

Dermal Pertaining to the skin. SN **cutaneous**.

Dermal irritation Skin reaction resulting from a single or multiple exposure to a physical or chemical entity at the same site, characterized by the presence of inflammation; it may result in cell death.

Dermatitis Inflammation of the skin: contact dermatitis is due to local exposure and may be caused by irritation, allergy, or infection.

Desensitization Suppression of sensitivity of an organism to an allergen to which the organism has been exposed previously.

Desquamation Shedding of an outer layer of skin in scales or shreds.

Detoxification
1. Process, or processes, of chemical modification that make a toxic molecule less toxic.
2. Treatment of patients suffering from poisoning in such a way as to promote physiological processes that reduce the probability or severity of harmful effects.

Diploid Chromosome state in which the chromosomes are present in homologous pairs. Normal human somatic (non-reproductive) cells are diploid (they have 46 chromosomes), whereas reproductive cells, with 23 chromosomes, are haploid. RT **haploid, meiosis, mitosis**.

Disease Literally, *dis-ease*, lack of *ease*; pathological condition that presents a group of symptoms peculiar to it and that establishes the condition as an abnormal entity different from other normal or pathological body states.

Disposition Natural tendency shown by an individual or group of individuals, including any tendency to acquisition of specific diseases, often due to hereditary factors.

Distributed source   See SN **area source**. RT **point source**.

Diuresis   Excretion of urine, especially in excess.

Diuretic   Agent that increases urine production. SN **micturitic**.

Dosage   Dose expressed as a function of the organism being dosed and time, for example mg/(kg body weight)/per day.

Dose   Total amount of a substance administered to, taken, or absorbed by an organism. NT **absolute lethal dose, cumulative median lethal dose, lethal dose, maximum tolerable dose, maximum tolerated dose, median effective dose, median lethal dose, median narcotic dose, minimum lethal dose, non-effective dose, organ dose, threshold dose, toxic dose**.

Dose–effect curve   Graph of the relation between dose and the magnitude of the biological change produced, measured in appropriate units. RT **concentration–effect curve**.

Dose–effect relationship   Association between dose and the magnitude of a continuously graded effect, in an individual or in a population or in experimental animals. RT **concentration–effect relationship**.

Dose-related effect   Situation in which the magnitude of a biological change is related to the dose. AN **non-dose-related effect**.

Dose–response curve   Graph of the relation between dose and the proportion of individuals in a population showing an all-or-none effect. RT **concentration–response curve, response**.

Dose–response relationship   Association between dose and the incidence of a defined biological effect in an exposed population. RT **concentration–response relationship, response**.

Drug   Any substance which when absorbed into a living organism may modify one or more of its functions. The term is generally accepted for a substance taken for a therapeutic purpose, but is also commonly used for abused substances. NT **medicine, pharmaceutical**.

Dysfunction   Abnormal, impaired, or incomplete functioning of an organism, organ, tissue, or cell.

Dysplasia Abnormal development of an organ or tissue identified by morphological examination.

Dyspnoea Difficult or laboured breathing.

Ecogenetics Study of the influence of hereditary factors on the effects of xenobiotics on individual organisms. PS **pharmacogenetics, toxicogenetics**. RT **polymorphism**.

Ecology Branch of biology that studies the interactions between living organisms and all factors (including other organisms) in their environment: such interactions encompass environmental factors that determine the distributions of living organisms.

Ecosystem Grouping of organisms (micro-organisms, plants, animals) interacting together, with and through their physical and chemical environments, to form a functional entity.

Ecotoxicology Study of the toxic effects of chemical and physical agents on all living organisms, especially on populations and communities within defined ecosystems; it includes transfer pathways of these agents and their interactions with the environment.

Ectoparasiticide Substance intended to kill parasites living on the exterior of the host.

Eczema Acute or chronic skin inflammation with erythema, papules, vesicles, pustules, scales, crusts, or scabs, alone or in combination, of varied aetiology.

Edema See SN **oedema**

Effective concentration (EC) Concentration of a substance that causes a defined magnitude of response in a given system: EC_{50} is the median concentration that causes 50% of maximal response. RT **lethal concentration**.

Effective dose (ED) Dose of a substance that causes a defined magnitude of response in a given system: ED_{50} is the median dose that causes 50% of maximal response. BT **dose. RT lethal dose**.

Elimination half-life or half-time Period taken for the plasma concentration of a substance to decrease by half. BT **biological half-life or half-time** $(t_{1/2})$.

Embryo Stage in the developing mammal at which the characteristic organs and organ systems are being formed:

1. For humans this involves the stages of development from the 2nd to the 8th week (inclusive of post-conception).
2. In birds, the stage of development from the fertilization of the ovum up to hatching.
3. In plants, the stage of development within the seed.

Embryotoxicity
1. Production by a substance of toxic effect in the progeny in the first period of pregnancy between conception and the fetal stage.
2. Any toxic effect on the conceptus as a result of prenatal exposure during the embryonic stages of development: these effects may include malformations and variations, malfunctions, altered growth, prenatal death, and altered postnatal function. RT **developmental toxicity, teratogenicity**.

Endemic   Present in a community or among a group of people; said of a disease prevailing continually in a region.

Endocrine   Pertaining to hormones or to the glands that secrete hormones directly into the bloodstream.

Endoplasmic reticulum   Intracellular complex of membranes in which proteins and lipids, as well as molecules for export, are synthesized and in which the biotransformation reactions of the monooxygenase enzyme systems occur: may be isolated as microsomes following cell fractionation procedures. RT **cytochrome P420, cytochrome P448, cytochrome P450, microsome, monooxygenase, phase 1 reaction**.

Endothelial   Pertaining to the layer of flat cells lining the inner surface of blood and lymphatic vessels, and the surface lining of serous and synovial membranes.

Enteritis   Intestinal inflammation.

Enterohepatic circulation   Cyclical process involving intestinal reabsorption of a substance that has been excreted through the bile followed by transfer back to the liver, making it available for biliary excretion again.

Environmental health impact assessment   Estimate of the adverse health effects or risks likely to follow from a proposed or expected environmental change or development.

Environmental impact assessment (EIA)   Appraisal of the possible environmental consequences of a past, ongoing, or planned action, resulting in the production of an environmental impact statement or 'finding of no significant impact (FONSI)'. RT **environmental impact statement**.

Environmental impact statement (EIS)   Report resulting from an environmental impact assessment. RT **environmental impact assessment**.

Environmental monitoring   Continuous or repeated measurement of agents in the environment to evaluate environmental exposure and possible damage by comparison with appropriate reference values based on knowledge of the probable relationship between ambient exposure and resultant adverse effects. RT **biological effect monitoring**, **biological monitoring**, **reference value**.

Environmental quality objective (EQO)   Overall state to be aimed for in a particular aspect of the natural environment, for example, 'water in an estuary such that shellfish populations survive in good health'. Unlike an environmental quality *standard*, the EQO is usually expressed in qualitative and not quantitative terms. RT **environmental quality standard**.

Environmental quality standard (EQS)   Amount, concentration, or mass concentration of a substance that should not be exceeded in an environmental system, often expressed as a time-weighted average measurement over a defined period. SN **ambient standard**. RT **limit value**.

Epidemiology   Study of the distribution and determinants of health-related states or events in populations and the application of this study to control of health problems.

Epigen/esis n., **-etic** adj. Changes in an organism brought about by alterations in the expression of genetic information without any change in the genome itself: the genotype is unaffected by such a change but the phenotype is altered. RT **mutation**, **phenotype**, **transformation**, **tumour**.

Epileptiform   Occurring in severe or sudden spasms, as in convulsion or epilepsy.

Epithelioma   Any tumour derived from epithelium. NT **carcinoma**.

Epithelium   Cells covering the internal and external surfaces of the body.

Epitope   Any part of a molecule that acts as an antigenic determinant: a macromolecule can contain many different epitopes each capable of stimulating production of a different specific antibody.

Erythema   Redness of the skin produced by congestion of the capillaries.

Estimated exposure concentration (EEC)   Measured or calculated amount or mass concentration of a substance to which an organism is likely to be exposed, considering exposure by all sources and routes.

Estimated exposure dose (EED)   Measured or calculated dose of a substance to which an organism is likely to be exposed, considering exposure by all sources and routes.

Estimated maximum daily intake (EMDI)   Prediction of the maximum daily intake of a residue of a potentially harmful agent based on assumptions of average food consumption per person and maximum residues in the edible portion of a commodity, corrected for the reduction or increase in residues resulting from preparation, cooking, or commercial processing. The EMDI is expressed in mg of residue per person.

Etiology   See **aetiology**

Eukaryote   Cell or organism with the genetic material packed in a membrane-surrounded structurally discrete nucleus and with well-developed cell organelles. The term includes all organisms except archaebacteria, eubacteria, and cyanobacteria (until recently classified as cyanophyta or blue-green algae). AN **prokaryote**.

European Inventory of Existing Commercial Chemical Substances (EINECS)   List of all substances supplied either singly or as components in preparations to persons in a Member State of the European Community on any occasion between 1 January 1971 and 18 September 1981.

Eutrophic   Describes a body of water with a high concentration of nutrient salts and a high or excessive rate of biological production.

Eutrophication   Adverse change in the chemical and biological status of a body of water following depletion of the oxygen content caused by decay of organic matter resulting from high primary production as a result of enhanced input of nutrients.

Excess lifetime risk   Additional or excess risk incurred over the lifetime of an individual by exposure to a toxic substance. BT **risk**. RT **hazard**.

Excipient   Any more or less inert substance added to a drug to give suitable consistency or form to the drug. RT **vehicle**.

Excretion   Discharge or elimination of an absorbed or endogenous substance, or of a waste product, and/or their metabolites, through some tissue of the body and its appearance in urine, faeces, or other products normally leaving the body. Excretion of substances from the body occurs mainly through the kidney and the gut. Volatile compounds may be largely eliminated by exhalation. Excretion by perspiration and through hair and nails may also occur. Excretion by the gastrointestinal tract may take place by various routes such as the bile, the shedding of intestinal cells, and transport through the intestinal mucosa. RT **clearance, elimination**.

Excretion rate Amount of substance (and/or its metabolites) or fraction that is excreted per unit time. It should be noted that according to this definition excretion does not include the passing of a substance through the intestine without absorption. When discussing the total amount of a substance in faeces (including the unabsorbed part), it is preferable to speak about faecal substance content (mol/kg) or mass content (kg/kg).

Exogenous Resulting from causes or derived from materials external to an organism. AN **endogenous**.

Exogenous substance See SN **xenobiotic**

Explant Living tissue removed from its normal environment and transferred to an artificial medium for growth.

Exposure
1. Concentration, amount, or intensity of a particular physical or chemical agent or environmental agent that reaches the target population, organism, organ, tissue, or cell, usually expressed in numerical terms of substance concentration, duration, and frequency (for chemical agents and micro-organisms) or intensity (for physical agents such as radiation).
2. Process by which a substance becomes available for absorption by the target population, organism, organ, tissue, or cell, by any route.

Exposure assessment Process of measuring or estimating concentration (or intensity), duration, and frequency of exposures to an agent present in the environment or, if estimating hypothetical exposures, that might arise from the release of a substance, or radionuclide, into the environment. RT **risk assessment**.

Exposure–effect relationship See NT **concentration–effect relationship, dose–effect relationship.**

Exposure limit General term defining an administrative substance concentration or intensity of exposure that should not be exceeded. RT **discharge limit**.

Exposure ratio In a case control study, value obtained by dividing the rate at which persons in the case group are exposed to the risk factor (or to the protective factor) by the rate at which persons in the control group are exposed to the risk factor (or to the protective factor) of interest.

Fecundity
1. Ability to produce offspring frequently and in large numbers.
2. In demography, the physiological ability to reproduce. PS **fertility**.

Feromone See **pheromone**

Fertility Ability to conceive and to produce offspring: for litter-bearing species the number of offspring per litter is used as a measure of fertility. Reduced fertility is sometimes referred to as subfertility. RT **fecundity**.

Fetotoxicity Toxicity to the fetus. RT **embryotoxicity, teratogenicity**.

Fetus (often incorrectly foetus) Young mammal within the uterus of the mother from the visible completion of characteristic organogenesis until birth: in humans, this period is usually defined as from the 3rd month after fertilization until birth (prior to this, the young mammal is referred to as an embryo). RT **embryo**.

Fibrosis Abnormal formation of fibrous tissue.

First-pass effect Biotransformation of a substance in the liver after absorption from the intestine and before it reaches the systemic circulation.

Fixed dose procedure Acute toxicity test in which a substance is tested initially at a small number (three or four) of predefined doses to identify which produces evident toxicity without lethality: the test may be repeated at one or more higher or lower, defined, discriminating doses to satisfy the criteria. NT **limit test**.

Fluorosis Adverse effects of fluoride, as in dental or skeletal fluorosis.

Foci (singular **focus**) Small groups of cells distinguishable, in appearance or histochemically, from the surrounding tissue: indicative of an early stage of a lesion that may lead to the formation of a neoplastic nodule.

Foetus See **fetus**

Follow-up study Investigation in which individuals or populations, selected on the basis of exposure to risk, who have received a specified preventive or therapeutic procedure, or who possess a certain characteristic, are followed to assess the outcome of exposure, the procedure, or effect of the characteristic, for example, occurrence of disease. SN **cohort study**.

Food chain Sequence of transfer of matter and energy in the form of food from organism to organism in ascending or descending trophic levels.

Food intolerance Physiologically based reproducible, unpleasant (adverse) reaction to a specific food or food ingredient that is not immunologically based. RT **food allergy**.

Food web Network of food chains.

Foreign substance (foreign compound) See SN **xenobiotic**

Frameshift mutation Point mutation involving either the deletion or insertion of one or two nucleotides in a gene: by the frameshift mutation, the normal reading frame used when decoding nucleotide triplets in the gene is altered.

Fumigant Substance that is vaporized in order to kill or repel pests.

Fungicide Substance intended to kill fungi.

Gamete Reproductive cell (either sperm or egg) containing a haploid set of chromosomes. RT **zygote**.

Gametocide Substance intended to kill gametes.

Gastrointestinal Pertaining or communicating with the stomach and intestine.

Gavage Administration of materials directly into the stomach by oesophageal intubation.

Gene Structurally a basic unit of hereditary material; an ordered sequence of nucleotide bases that encodes one polypeptide chain (following transcription to messenger RNA). SN **cistron**.

Gene amplification Production of extra copies of a chromosomal sequence found either as intra- or extra-chromosomal DNA; with respect to a plasmid, it refers to the increase in the number of plasmid copies per cell induced by a specific treatment of transformed cells.

Genetic polymorphism Condition in which a genetic character occurs in more than one form, resulting in the coexistence of more than one morphological type in a given population.

Genetic toxicology Study of substances that can produce adverse heritable changes.

Genome Complete set of chromosomal and extra-chromosomal genes of an organism, a cell, an organelle, or a virus: complete DNA component of an organism.

Genotoxicity Ability to cause damage to genetic material. Such damage may be mutagenic and/or carcinogenic.

Genotype Genetic constitution of an organism as revealed by genetic or molecular analysis; the complete set of genes, both dominant and recessive, possessed by a particular organism, cell, organelle, or virus.

Germ-free animal Animal grown under sterile conditions in the period of postnatal development: such animals are usually obtained by Caesarean operation and kept in special sterile boxes in which there are no viable micro-organisms (sterile air, food, and water are supplied). SN **axenic animal**. BT **gnotobiont**, **gnotobiote**.

Glomerular Pertaining to a tuft or cluster, as of a plexus of capillary blood vessels or nerve fibres, especially referring to the capillaries of the glomeruli of the kidney.

Gnotobiont See SN **gnotobiote**. NT **germ-free animal**.

Gnotobiota Specifically and entirely known microfauna and microflora of a specially reared laboratory animal. RT **gnotobiote**.

Gnotobiot/e n., **-ic** adj. Specially reared laboratory animal whose microflora and microfauna are specifically known in their entirety. NT **germ-free animal**.

Gonadotropic Pertaining to effects on sex glands and on the systems that regulate them.

Good laboratory practice (GLP) principles Fundamental rules incorporated in national regulations concerned with the process of effective organization and the conditions under which laboratory studies are properly planned, performed, monitored, recorded, and reported. RT **quality assurance**, **quality control**.

Graded effect Consequence that can be measured on a graded scale of intensity or severity and its magnitude related directly to the dose or concentration of the substance producing it. AN **all-or-none effect**, **quantal effect**. RT **stochastic effect**.

Granuloma Granular growth or tumour, usually of lymphoid and epithelial cells.

Guideline value Quantitative measure (a concentration or a number) of a constituent of an environmental medium that ensures aesthetically pleasing air, water, or food and does not result in a significant risk to the user.

Haematoma Localized accumulation of blood, usually clotted, in an organ, space, or tissue, due to a failure of the wall of a blood vessel.

Haematuria   Presence of blood in the urine.

Haemodialysis   Use of an artificial kidney to remove toxic compounds from the blood by passing it through a tube of semipermeable membrane. The tube is bathed in a dialysing solution to restore the normal chemical composition of the blood while permitting diffusion of toxic substances from the blood.

Haemoglobinuria   Presence of free haemoglobin in the urine.

Haemolysin   Substance that damages the membrane of erythrocytes causing the release of haemoglobin.

Haemolysis   Release of haemoglobin from erythrocytes, and its appearance in the plasma.

Haemoperfusion   Passing blood through a column of charcoal or adsorbent resin for the removal of drugs or toxins.

Haemosiderin   Iron-containing pigment that is formed from haemoglobin released during the disintegration of red blood cells and that accumulates in individuals who have ingested excess iron.

Half-life (half-time) $(t_{1/2})$ Time in which the concentration of a substance will be reduced by half, assuming a first-order elimination process or radioactive decay.

Haploid (monoploid)   State in which a cell contains only one set of chromosomes. RT **diploid, gamete, meiosis**.

Hapten   Low-molecular-mass molecule that contains an antigenic determinant (epitope) that may bind to a specific antibody but which is not itself antigenic unless complexed with an antigenic carrier such as a protein or cell; once bound it can cause the sensitization of lymphocytes, possibly leading to allergy or cell-mediated hypersensitivity. RT **allergy, antigen, antibody, cell-mediated hypersensitivity, epitope**.

Hazard   Set of inherent properties of a substance, mixture of substances, or a process involving substances that, under production, usage, or disposal conditions, make it capable of causing adverse effects to organisms or the environment, depending on the degree of exposure; in other words, it is a source of danger. RT **risk**.

Hazard assessment   Determination of factors controlling the likely effects of a hazard such as the dose–effect and dose–response relationships, variations in target susceptibility, and mechanism of toxicity. RT **exposure assessment, hazard evaluation, hazard identification, risk assessment, risk**

characterization, risk estimation, risk evaluation, risk identification, risk perception.

Hazard evaluation Establishment of a qualitative or quantitative relationship between hazard and benefit, involving the complex process of determining the significance of the identified hazard and balancing this against identifiable benefit: this may subsequently be developed into a risk evaluation. RT **exposure evaluation, hazard assessment, hazard identification, risk assessment, risk characterization, risk estimation, risk evaluation, risk identification, risk perception**.

Hazard identification Determination of substances of concern, their adverse effects, target populations, and conditions of exposure, taking into account toxicity data and knowledge of effects on human health, other organisms, and their environment.

Health
1. State of complete physical, mental, and social well-being, and not merely the absence of disease or infirmity.
2. State of dynamic balance in which an individual's or a group's capacity to cope with the circumstances of living is at an optimal level.
3. State characterized by anatomical, physiological, and psychological integrity; ability to perform personally valued family, work, and community roles; ability to deal with physical, biological, psychological, and social stress; a feeling of well-being; and freedom from the risk of disease and untimely death.

Health-based exposure limit Maximum concentration or intensity of exposure that can be tolerated without significant effect (based on only scientific and not economic evidence concerning exposure levels and associated health effects).

Health hazard Any factor or exposure that may adversely affect health.

Health surveillance Periodic medico-physiological examinations of exposed workers with the objective of protecting health and preventing occupationally related disease. RT **biological monitoring, biomarker, monitoring**.

Healthy worker effect Epidemiological phenomenon observed initially in studies of occupational diseases: workers usually exhibit lower overall disease and death rates than the general population, due to the fact that the old, severely ill, and disabled are ordinarily excluded from employment. Death rates in the general population may be inappropriate for comparison if this effect is not taken into account.

Hepatic Pertaining to the liver.

Hepatotoxic Poisonous to liver cells.

Herbicide Substance intended to kill plants.

Histology Study (usually microscopic) of the anatomy of tissues and their cellular and subcellular structure.

Histopathology Microscopic pathological study of the anatomy and cell structure of tissues in disease to reveal abnormal or adverse structural changes.

Homeostasis Normal, internal stability in an organism maintained by co-ordinated responses of the organ systems that automatically compensate for environmental changes.

Homology Degree of identity existing between the nucleotide sequences of two related but not complementary DNA or RNA molecules; 70% homology means that on average 70 out of every 100 nucleotides are identical in a given sequence. The same term is used in comparing the amino acid sequences of related proteins.

Hormesis Stimulatory effect of small doses of a potentially toxic substance that is inhibitory in larger doses.

Hormone Substance formed in one organ or part of the body and carried in the blood to another organ or part where it selectively alters functional activity.

Human ecology Inter-relationship between humans and the entire environment—physical, biological, socio-economic, and cultural—including the inter-relationships between individual humans or groups of humans and other human groups or groups of other species.

Hygiene Science of health and its preservation.

Hyper- Prefix meaning above or excessive: when used with the suffix '-aemia' refers to blood and with the suffix '-uria' refers to urine, for example 'hyperbilirubinaemia'.

Hyperaemia Excessive amount of blood in any part of the body.

Hyperalimentation Ingestion or administration of nutrients in excess of optimal amounts.

Hyperbilirubinaemia Excessive concentration of bilirubin in the blood.

Hypercalcaemia Excessive concentration of calcium in the blood.

Hyperglycaemia   Excessive concentration of glucose in the blood.

Hyperkalaemia   Excessive concentration of potassium in the blood.

Hypernatraemia   Excessive concentration of sodium in the blood.

Hyperparathyroidism   Abnormally increased parathyroid gland activity that affects, and is affected by, plasma calcium concentration.

Hyperplasia   Abnormal multiplication or increase in the number of normal cells in a tissue or organ. RT **hypertrophy, neoplasia**.

Hypersensitivity   State in which an individual reacts with allergic effects following exposure to a certain substance (allergen) after having been exposed previously to the same substance. PS **allergy**. RT **cell-mediated hypersensitivity, sensitization**.

Hypersusceptibility   Excessive reaction following exposure to a given amount or concentration of a substance as compared with most other exposed subjects. RT **idiosyncrasy**.

Hypertension   Persistently high blood pressure in the arteries or in a circuit, for example pulmonary hypertension or hepatic portal hypertension.

Hypertrophy   Excessive growth in bulk of a tissue or organ through increase in size but not in number of the constituent cells. RT **hyperplasia**.

Hypervitaminosis   Condition resulting from the ingestion of an excess of one or more vitamins.

Hypo-   Prefix meaning under, deficient: when used with the suffix '-aemia' refers to blood and with the suffix '-uria' refers to urine, for example 'hypocalcaemia'.

Hypocalcaemia   Abnormally low calcium concentration in the blood.

Hypokalaemia   Abnormally low potassium concentration in the blood.

Hyponatraemia   Abnormally low sodium concentration in the blood.

Hypovolaemic   Pertaining to an abnormally decreased volume of circulating fluid (plasma) in the body.

Hypoxaemia   Deficient oxygenation of the blood.

Hypoxia
1. Abnormally low oxygen content or tension.
2. Deficiency of oxygen in the inspired air, in blood, or in tissues, short of anoxia.

Iatrogenic   Any adverse condition resulting from medical treatment. NT **nosocomial**.

Icterus   Excess of bile pigment in the blood and consequent deposition and retention of bile pigment in the skin and the sclera. RT **hyperbilirubinaemia**, **jaundice**.

Idiosyncrasy   Genetically based, unusually high sensitivity of an organism to the effect of certain substances. RT **hypersusceptibility**, **pharmacogenetics**.

Immune complex   Product of an antigen–antibody reaction that may also contain components of the complement system.

Immune response   Selective reaction of the body to substances that are foreign to it, or that the immune system identifies as foreign, shown by the production of antibodies and antibody-bearing cells or by a cell-mediated hypersensitivity reaction. RT **antibody**, **autoimmune disease**, **cell-mediated hypersensitivity**.

Immunochemistry   Study of biochemical and molecular aspects of immunology, especially the nature of antibodies and antigens, and their interactions.

Immunogen   See SN **antigen**

Immunoglobulin   Family of closely related glycoproteins capable of acting as antibodies and present in plasma and tissue fluids; immunoglobulin E is the source of antibody in many hypersensitivity (allergic) reactions. RT **allergy**, **antibody**, **hypersensitivity**.

Immunoglobulin E-mediated hypersensitivity   State in which an individual reacts with allergic effects caused fundamentally by the reaction of antigen-specific immunoglobulin E following exposure to a certain substance (allergen) after having been exposed previously to the same substance. RT **allergy**, **antibody**, **antigen**, **cell-mediated hypersensitivity**, **hypersensitivity**, **immunoglobulin**.

Immunopotentiation   Enhancement of the capacity of the immune system to produce an effective response.

Immunosuppression   Reduction in the functional capacity of the immune response; maybe due to:

1. Inhibition of the normal response of the immune system to an antigen.
2. Prevention, by chemical or biological means, of the production of an antibody to an antigen by inhibition of the processes of transcription, translation, or formation of tertiary structure.

Immunosurveillance   Mechanisms by which the immune system is able to recognize and destroy malignant cells before the formation of an overt tumour.

Immunotoxic   Poisonous to the immune system.

Incidence   Number of occurrences of illness commencing, or of persons falling ill, during a given period in a specific population: usually expressed as a rate, the denominator being the average number of persons in the specified population during a defined period or the estimated number of persons at the mid-point of that period. The basic distinction between 'incidence' and 'prevalence' is that whereas incidence refers only to new cases, prevalence refers to all cases, irrespective of whether they are new or old. When the terms incidence and prevalence are used, it should be stated clearly whether the data represent the numbers of instances of the disease recorded or the numbers of persons ill.

Incidence rate   Measure of the frequency with which new events occur in a population. Value obtained by dividing the number of new events that occur in a defined period by the population at risk of experiencing the event during this period, sometimes expressed as person-time.

Indirect exposure
1 Exposure to a substance in a medium or vehicle other than the one originally receiving the substance.
2. Exposure of people to a substance by contact with a person directly exposed.
RT **bystander exposure**, **para-occupational exposure**.

Individual protective device (IPD)   Device for individual use for protection of the whole body, eyes, respiratory pathways, or skin of workers against hazardous and harmful production factors. SN **personal protective device (PPD)**, **personal protective equipment (PPE)**.

Individual risk   Probability that an individual person will experience an adverse effect.

Inducer   Substance that causes induction. RT **induction**.

Induction   Increase in the rate of synthesis of an enzyme in response to the action of an inducer or environmental conditions: often the substrate of the induced enzyme or a structurally similar substance (gratuitous inducer) that is not metabolized.

Induction period   Time from the onset of exposure to the appearance of signs of disease. SN **latent period**.

Inhibitory concentration (IC)   Concentration of a substance that causes a defined inhibition of a given system: IC_{50} is the median concentration that causes 50% inhibition. RT **effective concentration**, **lethal concentration**.

Inhibitory dose (ID)   Dose of a substance that causes a defined inhibition of a given system: ID_{50} is the median dose that causes 50% inhibition. RT **effective dose**, **lethal dose**.

Initiator
1. Agent that induces a change in a chromosome or gene that leads to the induction of tumours after a second agent, called a promoter, is administered to the tissue. RT **promoter**.
2. Substance that starts a chain reaction; an initiator is consumed in a reaction, in contrast to a catalyst.

Insecticide   Substance intended to kill insects.

Interstitial pneumonia   Chronic form of pneumonia involving increase of the interstitial tissue and decrease of the functional lung tissue.

Intervention study   Epidemiological investigation designed to test a hypothesized cause–effect relationship by modifying a supposed causal factor in a population.

Intestinal reabsorption   Absorption further down the intestinal tract of a substance or substances that have been absorbed before and subsequently excreted into the intestinal tract, usually through the bile.

Intoxication
1. Poisoning: pathological process with clinical signs and symptoms caused by a substance of exogenous or endogenous origin. RT **exogenous**, **endogenous**.
2. Drunkenness following consumption of beverages containing ethanol or other compounds affecting the central nervous system.

In vitro   In glass, referring to a study in the laboratory usually involving isolated organ, tissue, cell, or biochemical systems. AN *in vivo*.

In vivo   In the living body, referring to a study performed on a living organism. AN *in vitro*.

Ionizing radiation   Any radiation consisting of directly or indirectly ionizing particles or a mixture of both, or photons with energy higher than the energy of photons of ultraviolet light, or a mixture of both such particles and photons.

Irritant
1. n. Substance that causes inflammation following immediate, prolonged, or repeated contact with skin, mucous membrane, or other biological material. A substance capable of causing inflammation on first contact is called a primary irritant.
2. adj. Causing inflammation following immediate, prolonged, or repeated contact with skin, mucous membrane, or other tissues.

Ischaemia   Local deficiency of blood supply and hence oxygen to an organ or tissue owing to constriction of the blood vessels or to obstruction.

Itai-itai disease   Illness observed in Japan possibly resulting from the ingestion of cadmium-contaminated rice: damage occurs to the renal and skeleto-articular systems, the latter being very painful ('itai' means pain in Japanese).

Jaundice   Pathological condition characterized by deposition of bile pigment in the skin and mucous membranes, including the conjunctivae, resulting in a yellow appearance of the patient or animal. RT **hyperbilirubinaemia, icterus**.

Lachrymator   See **lacrimator**

Lacrimator   Substance that irritates the eyes and causes the production of tears or increases the flow of tears.

Larvicide   Substance intended to kill larvae.

Laryngospasm   Reflex spasmodic closure of the sphincter of the larynx, particularly the glottic sphincter.

Larynx   Main organ of voice production; the part of the respiratory tract between the pharynx and the trachea.

Latent effect   Consequence occurring after a delay following the end of exposure to a potentially toxic substance or other harmful environmental factor.

Latent period   Delay between exposure to a disease-causing agent and the appearance of manifestations of the disease: also defined as the period from disease initiation to disease detection. SN **latency**.

Lavage   Irrigation or washing out of a hollow organ or cavity such as the stomach, intestine, or lungs.

Laxative Substance that causes evacuation of the intestinal contents. SN **cathartic, purgative**.

Lesion
1. Area of pathologically altered tissue.
2. Injury or wound.
3. Infected patch of skin.

Lethal concentration Concentration of a potentially toxic substance in an environmental medium that causes death following a certain period of exposure (denoted by LC). RT **effective concentration, lethal dose**.

Lethal dose Amount of a substance or physical agent (radiation) that causes death when taken into the body by a single absorption (denoted by LD). RT **effective dose, lethal concentration**.

Lethal synthesis Metabolic formation of a highly toxic compound from one that is relatively non-toxic (bioactivation), often leading to death of affected cells. SN **suicide metabolism**.

Leukaemia Progressive, malignant disease of the blood-forming organs, characterized by distorted proliferation and development of leucocytes and their precursors in the bone marrow and blood .

Leukopenia Reduced concentration of leucocytes in the blood.

Lg K_{ow} See SN **lg P_{ow}**

LgP_{ow} Logarithm of base 10 of the partition coefficient of a substance between octan-1–ol and water: as an empirical measure for lipophilicity used for calculating bioaccumulation, fish toxicity, membrane adsorption, and penetration *etc*. RT **lipophilicity, octanol–water partition coefficient, partition coefficient**. SN **lg K_{ow}**.

Limacide Substance intended to kill mollusca including the gastropod mollusc, *Limax.*

Limit test Acute toxicity test in which, if no ill-effects occur at a pre-selected maximum dose, no further testing at greater exposure levels is required. RT **fixed dose procedure**.

Limit value (LV) Limit concentration at or below which Member States of the European Community must set their environmental quality standard and emission standard for a particular substance according to Community Directives. NT **threshold limit value**.

Linearized multistage model Sequence of steps in which (a) a multistage model is fitted to tumour incidence data; (b) the maximum linear term consistent with the data is calculated; (c) the low-dose slope of the dose–response function is equated to the coefficient of the maximum linear term; and (d) the resulting slope is then equated to the upper bound of potency. BT **multistage model**.

Liposome   Originally a lipid droplet in the endoplasmic reticulum of a fatty liver. Now applied to an artificially formed lipid droplet, small enough to form a relatively stable suspension in aqueous media and with potential use in drug delivery.

Local effect   Circumscribed change occurring at the site of contact between an organism and a toxicant. RT **systemic effect**.

Logit transformation   Mathematical transformation that relates response to a stated dose or concentration of a toxicant to the response in the absence of the toxicant by the formula:

$$\text{Logit} = \lg \left[B/(B_0 - B) \right]$$

where B is the response to the stated dose or concentration and B_0 is the response in the absence of the toxicant. Plotting the logit function against the logarithm of base 10 of the dose or concentration usually gives a linear relationship.

Lowest observed adverse effect level (LOAEL)   Lowest concentration or amount of a substance, found by experiment or observation, that causes an adverse alteration of morphology, functional capacity, growth, development, or life span of a target organism distinguishable from normal (control) organisms of the same species and strain under defined conditions of exposure. RT **adverse effect**, **lowest observed effect level**, **no observed adverse effect level**, **no observed effect level**.

Lowest observed effect level (LOEL)   Lowest concentration or amount of a substance, found by experiment or observation, that causes any alteration in morphology, functional capacity, growth, development, or life span of target organisms distinguishable from normal (control) organisms of the same species and strain under the same defined conditions of exposure. **RT adverse effect**, **lowest observed adverse effect level**, **no observed adverse effect level**, **no observed effect level**.

Lymphocyte   Animal cell that interacts with a foreign substance or organism, or one which it identifies as foreign, and initiates an immune response against the substance or organism. There are two groups of lymphocytes, B-lymphocytes and T-lymphocytes. NT **B-lymphocyte**, **immune response**, **T-lymphocyte**.

Lymphoma   General term comprising tumours and conditions allied to tumours arising from some or all of the cells of lymphoid tissue.

Lysimeter Laboratory column of selected representative soil or a protected monolith of undisturbed field soil with which it is possible to sample and monitor the movement of water and substances.

Lysosome Membrane-bound cytoplasmic organelle containing hydrolytic enzymes.

Macrophage Large (10–20 mm diameter) amoeboid and phagocytic cell found in many tissues, especially in areas of inflammation; macrophages are derived from blood monocytes and play an important role in host defence mechanisms.

Mainstream smoke (tobacco smoking) Smoke that is inhaled. RT **sidestream smoke**.

Malaise Vague feeling of bodily discomfort.

Malignancy Population of cells showing both uncontrolled growth and a tendency to invade and destroy other tissues; a malignancy is life-threatening. RT **cancer, metastasis, tumour**.

Malignant
  1. Tending to become progressively worse and to result in death if not treated.
  2. In cancer, cells showing both uncontrolled growth and a tendency to invade and destroy other tissues. AN **benign**.

Mania Emotional disorder (mental illness) characterized by an expansive and elated state (euphoria), rapid speech, flight of ideas, decreased need for sleep, distractability, grandiosity, poor judgement, and increased motor activity.

Margin of exposure (MOE), margin of safety (MOS) Ratio of the no observed adverse effect level (NOAEL) to the theoretical or estimated exposure dose (EED) or concentration (EEC). RT **therapeutic index**.

Maximum allowable (admissible, acceptable) concentration (MAC) Regulatory value defining the concentration that if inhaled daily (in the case of work people for 8 hours with a working week of 40 hours; in the case of the general population 24 hours) does not, in the present state of knowledge, appear capable of causing appreciable harm, however long delayed, during the working life or during subsequent life or in subsequent generations. RT **permissible exposure limit, threshold limit value**.

Maximum exposure limit (MEL) Occupational exposure limit legally defined in the UK under COSHH (Control of Substances Hazardous to Health, 1988 Regulations) as the maximum concentration of an airborne substance, averaged over a reference period, to which employees may be exposed by inhalation under any circumstances, and set on the advice of the Health and Safety Commission Advisory Committee on Toxic Substances. RT **ceiling value**.

Maximum permissible concentration (MPC) See SN **maximum allowable concentration**

Maximum permissible daily dose Maximum daily dose of substance whose penetration into a human body during a lifetime will not cause diseases or health hazards that can be detected by current investigation methods and will not adversely affect future generations.

Maximum permissible level (MPL) Level, usually a combination of time and concentration, beyond which any exposure of humans to a chemical or physical agent in their immediate environment is unsafe. RT **maximum allowable concentration**.

Maximum tolerable concentration (MTC) Highest concentration of a substance in an environmental medium that does not cause death of test organisms or species (denoted by LC_0).

Maximum tolerated dose (MTD) High dose used in chronic toxicity testing that is expected on the basis of an adequate subchronic study to produce limited toxicity when administered for the duration of the test period. It should not induce (a) overt toxicity, for example appreciable death of cells or organ dysfunction, or (b) toxic manifestations that are predicted materially to reduce the life span of the animals except as the result of neoplastic development or (c) 10% or greater retardation of body weight gain as compared with control animals. In some studies, toxicity that could interfere with a carcinogenic effect is specifically excluded from consideration.

Median effective concentration (EC_{50}) Statistically derived concentration of a substance in an environmental medium expected to produce a certain effect in 50% of test organisms in a given population under a defined set of conditions.

Median effective dose (ED_{50}) Statistically derived dose of a chemical or physical agent (radiation) expected to produce a certain effect in 50% of test organisms in a given population or to produce a half-maximal effect in a biological system under a defined set of conditions.

Median lethal concentration (LC_{50}) Statistically derived concentration of a substance in an environmental medium expected to kill 50% of organisms in a given population under a defined set of conditions.

Median lethal dose (LD_{50}) Statistically derived dose of a chemical or physical agent (radiation) expected to kill 50% of organisms in a given population under a defined set of conditions.

Median lethal time (TL$_{50}$)   Statistically derived average time interval during which 50% of a given population may be expected to die following acute administration of a chemical or physical agent (radiation) at a given concentration under a defined set of conditions.

Meiosis
1. Process of 'reductive' cell division, occurring in the production of gametes, by means of which each daughter nucleus receives half the number of chromosomes characteristic of the somatic cells of the species. RT **chromosome, diploid, gamete, haploid**.
2. See **miosis**.

Mesocosm   See RT **microcosm**.

Mesothelioma   Malignant tumour of the mesothelium of the pleura, pericardium, or peritoneum; this tumour may be caused by exposure to asbestos fibres and some other fibres. BT **tumour**. RT **malignant**.

Metabolic activation   Biotransformation of a substance of relatively low toxicity to a more toxic derivative. BT **activation, biotransformation**. NT **lethal synthesis**. SN **bioactivation**.

Metabolic half-life (half-time)   Time required for one-half of the quantity of a substance in the body to be metabolically transformed into a derivative or to be eliminated. RT **clearance, elimination**.

Metabolic model   Analysis and theoretical reconstruction of the way in which the body deals with a specific substance, showing the proportion of the intake that is absorbed, the proportion that is stored and in what tissues, the rate of breakdown in the body and the subsequent fate of the metabolic products, and the rate at which it is eliminated by different organs as unchanged substance or metabolites.

Metabolic transformation   Biochemical transformation of a substance that takes place within an organism. SN **biotransformation**.

Metabolism   Sum total of all physical and chemical processes that take place within an organism; in a narrower sense, the physical and chemical changes that take place in a given substance within an organism. It includes the uptake and distribution within the body of chemical compounds, the changes (biotransformation) undergone by such substances, and the elimination of the compounds and their metabolites. RT **biotransformation**.

Metabolite Any intermediate or product resulting from metabolism. RT **biotransformation**.

Metaplasia Abnormal transformation of an adult, fully differentiated tissue of one kind into a differentiated tissue of another kind. RT **hyperplasia**, **neoplasia**.

Metastasis
1. Movement of bacteria or body cells, especially cancer cells, from one part of the body to another, resulting in change in location of a disease or of its symptoms from one part of the body to another.
2. Growth of pathogenic micro-organisms or of abnormal cells distant from the site of their origin in the body.

Methaemoglobinaemia Presence of methaemoglobin (oxidized haemoglobin) in the blood in greater than normal proportion.

Methaemoglobin-forming substance Substance capable of oxidizing directly or indirectly the iron(II) in haemoglobin to iron(III) to form methaemoglobin, a derivative of haemoglobin that cannot transport oxygen.

Microalbuminuria Chronic presence of albumin in slight excess in urine.

Microcosm Artificial test system that simulates major characteristics of the natural environment for the purposes of ecotoxicological assessment: such a system would commonly have a terrestrial phase, with substrate, plants, and herbivores, and an aquatic phase, with vertebrates, invertebrates, and plankton. The term 'mesocosm' implies a more complex and larger system than the term 'microcosm' but the distinction is not clearly defined. SN **experimental model ecosystem**.

Microsome Artefactual spherical particle, not present in the living cell, derived from pieces of the endoplasmic reticulum present in homogenates of tissues or cells. Microsomes sediment from such homogenates when centrifuged at $100\,000\ g$ and higher: the microsomal fraction obtained in this way is often used as a source of monooxygenase enzymes. RT **cytochrome P420**, **cytochrome P448**, **cytochrome P450**, **endoplasmic reticulum**, **monooxygenase**, **phase 1 reaction**.

Micturitic See SN **diuretic**

Minamata disease Neurological disease caused by ingestion of methylmercury-contaminated fish, first seen at Minamata Bay in Japan.

Mineralization Complete conversion of organic substances to inorganic derivatives. See SN **complete mineralization**.

Minimum lethal concentration (LC$_{min}$) Lowest concentration of a toxic substance in an environmental medium that kills individual organisms or test species under a defined set of conditions. SN **lowest lethal concentration found**.

Minimum lethal dose (LD$_{min}$) Lowest amount of a substance that, when introduced into the body, may cause death to individual species of test animals under a defined set of conditions.

Miosis Abnormal contraction of the pupil of the eye to less than 2 mm. Alternative spelling (obsolete): meiosis.

Mitochondri/on (pl. **-a**) Eukaryote cytoplasmic organelle that is bounded by an outer membrane and an inner membrane; the inner membrane has folds called cristae that are the centre of ATP (adenosine 5'-triphosphate) synthesis in oxidative phosphorylation in the animal cell and supplement ATP synthesis by the chloroplasts in photosynthetic cells. The mitochondrial matrix within the inner membrane contains ribosomes, many oxidative enzymes, and a circular DNA molecule that carries the genetic information for a number of these enzymes.

Mitogen Substance that induces lymphocyte transformation or, more generally, mitosis and cell proliferation. RT **transformation**.

Mitosis Process by which a cell nucleus divides into two daughter nuclei, each having the same genetic complement as the parent cell: nuclear division is usually followed by cell division.

Mixed function oxidase See SN **monooxygenase**

Modifying factor (MF) As used by the US Environmental Protection Agency, uncertainty factor that is greater than zero and less than, or equal to 10; the magnitude of the factor depends upon the professional assessment of scientific uncertainties of a study or database not explicitly treated with the standard uncertainty factors (for example the completeness of the overall database and the number of animals tested); the default value for the factor is 1. BT **assessment factor, safety factor, uncertainty factor**.

Molluscicide Substance intended to kill molluscs. SN **limacide.**

Monoclonal Pertaining to a specific protein from a single clone of cells, all molecules of this protein being the same.

Monoclonal antibody Antibody produced by cloned cells derived from a single lymphocyte. BT **antibody**. RT **polyclonal antibody**.

Monooxygenase   Enzyme that catalyses reactions between an organic compound and molecular oxygen in which one atom of the oxygen molecule is incorporated into the organic compound and one atom is reduced to water; involved in the metabolism of many natural and foreign compounds giving both unreactive products and products of different or increased toxicity from that of the parent compound. Such enzymes are the main catalysts of phase 1 reactions in the metabolism of xenobiotics by the endoplasmic reticulum or by preparations of microsomes. SN **mixed function oxidase**. RT **cytochrome P420, cytochrome P448, cytochrome P450, endoplasmic reticulum, microsome, phase 1 reaction**.

Morbidity   Any departure, subjective or objective, from a state of physiological or psychological well-being: in this sense, 'sickness', 'illness', and 'morbid condition' are similarly defined and synonymous. The WHO Expert Committee on Health Statistics noted in its Sixth Report (1959) that morbidity could be measured in terms of three units:
1. Proportion of persons who were ill.
2. The illnesses (periods or spells of illnesses) that these persons experienced.
3. The duration (days, weeks, *etc.*) of these illnesses. NT **disease**.

Morbidity rate   Term used loosely to refer to incidence or prevalence rates of disease.

Morbidity survey   Method for the estimation of the prevalence and/or incidence of a disease or diseases in a population. A morbidity survey is usually designed simply to ascertain the facts as to disease distribution, and not to test a hypothesis.

Mortality study   Investigation dealing with death rates or proportion of deaths attributed to specific causes as a measure of response.

Multigeneration study
1. Toxicity test in which two to three generations of the test organism are exposed to the substance being assessed.
2. Toxicity test in which only one generation is exposed and effects on subsequent generations are assessed.

Multistage model   Dose–response model for cancer death estimation of the form

$$P(d) = 1 - \exp[-(q_0 + q_1 d^1 + q_2 d^2 + \ldots + q_{(k)} d^k)]$$

where $P(d)$ is the probability of cancer death from a continuous dose rate, d, the q's are constants, and k is the number of dose groups (or, if less than the number of dose groups, k is the number of biological stages believed to be required in the carcinogenesis process). With the multistage model, it is assumed that cancer is initiated by cell mutations in a finite series of steps. A one-stage model is equivalent to a one-hit model.

Mutagen Any substance that can induce heritable changes (mutations) of the genotype in a cell as a consequence of alterations or loss of genes or chromosomes (or parts thereof).

Mutagenesis Introduction of heritable changes (mutations) of the genotype in a cell as a consequence of alterations or loss of genes or chromosomes (or parts thereof).

Mutagenicity Ability of a physical, chemical, or biological agent to induce heritable changes (mutations) in the genotype in a cell as a consequence of alterations or loss of genes or chromosomes (or parts thereof).

Mutation Any relatively stable heritable change in genetic material that may be a chemical transformation of an individual gene (gene or point mutation), altering its function, or a rearrangement, gain, or loss of part of a chromosome, that may be microscopically visible (chromosomal mutation). Mutation can be either germinal and inherited by subsequent generations, or somatic and passed through cell lineage by cell division. RT **chromosome, clastogenesis, gene, genotoxicity**.

Myasthenia Muscular weakness.

Mycotoxin Toxin produced by a fungus.

Myelosuppression Reduction of bone marrow activity leading to a lower concentration of platelets, red cells, and white cells in the blood.

Narcotic
1. Non-specific usage: an agent that produces insensibility or stupor.
2. Specific usage: an opioid; any natural or synthetic drug that has morphine-like actions.

Natriuretic Substance increasing the rate of excretion of sodium ion in the urine.

Necropsy See SN **autopsy**. RT **biopsy**

Necrosis
1. Mass death of areas of tissue or bone surrounded by healthy areas.
2. Morphological changes that follow cell death, characterized most frequently by nuclear changes.

Negligible risk
1. Probability of adverse effects occurring that can reasonably be described as trivial.
2. Probability of adverse effects occurring that is so low that it cannot be reduced appreciably by increased regulation or investment of resources. RT **acceptable risk, risk *de minimis*.**

Nematocide   Substance intended to kill nematodes.

Neonat/e n., **-al** adj. Infant during the first 4 weeks of postnatal life; for statistical purposes some scientists have defined the period as the first 7 days.

Neoplas/ia, -m   New and abnormal formation of tissue as a tumour or growth by cell proliferation that is faster than normal and continues after the initial stimulus that initiated the proliferation has ceased. PS **tumour**. RT **hyperplasia, metaplasia**.

Nephritis   Inflammation of the kidney, leading to kidney failure, usually accompanied by proteinuria, haematuria, oedema, and hypertension.

Nephrotoxic   Chemically harmful to the cells of the kidney.

Neural   Pertaining to a nerve or to the nerves.

Neuron(e)   Nerve cell, the morphological and functional unit of the central and peripheral nervous systems.

Neuropathy   Any disease of the central or peripheral nervous system.

Neurotoxic/ adj., **-ity**   n. Able to produce chemically an adverse effect on the nervous system: such effects may be subdivided into two types.
1. Central nervous system effects (including transient effects on mood or performance and pre-senile dementia such as Alzheimer's disease).
2. Peripheral nervous system effects (such as the inhibitory effects of organo-phosphorus compounds on synaptic transmission).

***N*-octanol–water partition coefficient**   See SN **octanol–water partition coefficient**

No effect level (NEL)   Maximum dose (of a substance) that produces no detectable changes under defined conditions of exposure. At present, this term tends to be substituted by no observed adverse effect level (NOAEL) or no observed effect level (NOEL). RT **adverse effect, no observed adverse effect level (NOAEL), no observed effect level (NOEL)**.

Non-target organism Organism affected by a pesticide although not the intended object of its use.

No observed adverse effect level (NOAEL) Greatest concentration or amount of a substance, found by experiment or observation, that causes no detectable adverse alteration of morphology, functional capacity, growth, development, or life span of the target organism under defined conditions of exposure. RT **adverse effect**.

No observed effect level (NOEL) Greatest concentration or amount of a substance, found by experiment or observation, that causes no alterations of morphology, functional capacity, growth, development, or life span of target organisms distinguishable from those observed in normal (control) organisms of the same species and strain under the same defined conditions of exposure. RT **adverse effect**.

No response level Maximum dose of a substance at which no specified response is observed in a defined population and under defined conditions of exposure.

Nosocomial Associated with a hospital or infirmary, especially used of diseases that may result from treatment in such an institution. BT **iatrogenic**.

Nuisance threshold Lowest concentration of an air pollutant that can be considered objectionable. RT **odour threshold, pollutant**.

Nystagmus Involuntary, rapid, rhythmic movement (horizontal, vertical, rotary, mixed) of the eyeball, usually caused by a disorder of the labyrinth of the inner ear or a malfunction of the central nervous system.

Occupational exposure limit (OEL) Regulatory level of exposure to substances, intensities of radiation *etc.*, or other conditions, specified appropriately in relevant government legislation or related codes of practice.

Occupational exposure standard (OES)
1. Level of exposure to substances, intensities of radiation *etc.* or other conditions considered to represent specified good practice and a realistic criterion for the control of exposure by appropriate plant design, engineering controls, and, if necessary, the addition and use of personal protective clothing.
2. In the UK, health-based exposure limit defined under COSHH (Control of Substances Hazardous to Health) Regulations as the concentration of any airborne substance, averaged over a reference period, at which, according to current knowledge, there is no evidence that it is likely to be injurious to employees, if they are exposed by inhalation, day after day, to that

concentration, and set on the advice of the Health and Safety Executive Advisory Committee on Toxic Substances.

Occupational hygiene   Identification, assessment, and control of physico-chemical and biological factors in the workplace that may affect the health or well-being of those at work and in the surrounding community.

Octanol–water partition coefficient (P_{ow}, K_{ow})   Measure of lipophilicity by determination of the equilibrium distribution between octan-1-ol and water, used in the assessment of biological fate and transport of organic chemicals.

Ocular   Pertaining to the eye.

Odds   Ratio of the probability of occurrence of an event to that of non-occurrence, or the ratio of the probability that something is so, to the probability that it is not so.

Odds ratio   Quotient obtained by dividing one set of odds by another. The term 'odds' or 'odds ratio' is defined differently according to the situation under discussion. Consider the following notation for the distribution of a binary exposure and a disease in a population or a sample.

	Exposed	*Non-exposed*
Disease	a	b
No disease	c	d

The odds ratio (cross-product ratio) is $ad/(bc)$.

Notes:

1. The exposure-odds ratio for a set of case control data is the ratio of the odds in favour of exposure among the cases (a/b) to the odds in favour of exposure among non-cases (c/d). This reduces to $ad/(bc)$. With incident cases, unbiased subject selection, and a 'rare' disease (say under 2% cumulative incidence rate over the study period), $ad/(bc)$ is an approximate estimate of the risk ratio. With incident cases, unbiased subject selection, and density sampling of controls, $ad/(bc)$ is an estimate of the ratio of the person-time incidence rates (forces of morbidity) in the exposed and unexposed. No rarity assumption is required.

2. The disease-odds (rate-odds) ratio for a cohort or cross-section is the ratio of the odds in favour of disease among the exposed population (a/c) to the odds in favour of disease among the unexposed (b/d). This reduces to $ad/(bc)$ and hence is equal to the exposure-odds ratio for the cohort or cross-section.

3. The prevalence-odds refers to an odds ratio derived cross-sectionally, as, for example, an odds ratio derived from studies of prevalent (rather than incident) cases.

4. The risk-odds ratio is the ratio of the odds in favour of getting disease, if exposed, to the odds in favour of getting disease if not exposed. The odds

ratio derived from a cohort study is an estimate. SN **cross-product ratio, relative odds**.

Odour threshold In principle, the lowest concentration of an odorant that can be detected by a human being: in practice, a panel of 'sniffers' is used, and the threshold taken as the concentration at which 50% of the panel can detect the odorant (although some workers have also used 100% thresholds).

Oedema Presence of abnormally large amounts of fluid in intercellular spaces of body tissues.

Oliguria Excretion of a diminished amount of urine in relation to fluid intake.

Oncogene Gene that can cause neoplastic transformation of a cell; oncogenes are slightly changed equivalents of normal genes known as proto-oncogenes. RT **transformation**.

Oncogenesis Production or causation of tumours.

One-hit model Dose–response model of the form

$$P(d) = 1 - \exp(-bd)$$

where $P(d)$ is the probability of cancer death from a continuous dose rate (d) and b is a constant. The one-hit model is based on the concept that a tumour can be induced after a single susceptible target or receptor has been exposed to a single effective dose unit of an agent.

Onycholysis Loosening or detachment of the nail from the nailbed following some destructive process.

Oogenesis Process of formation of the ovum (plural ova), the female gamete.

Operon Complete unit of gene expression and regulation, including structural genes, regulator gene(s), and control elements in DNA recognized by regulator gene product(s).

Ophthalmic Pertaining to the eye.

Organelle Microstructure or separated compartment within a cell that has a specialized function, *e.g.* ribosome, peroxisome, lysosome, Golgi apparatus, mitochondrion, nucleolus, nucleus.

Organic carbon partition coefficient (K_{oc}) Measure of the tendency for organic substances to be adsorbed by soil and sediment, expressed as:

$$K_{oc} = \frac{(\text{mg of substance adsorbed})/(\text{kg of organic carbon})}{(\text{mg of substance dissolved})/(\text{litre of solution})}$$

The K_{oc} is substance-specific and is largely independent of soil properties.

Organoleptic   Involving an organ, especially a sense organ as of taste, smell, or sight.

Osteo-   Prefix meaning pertaining to bone.

Osteodystrophy   Abnormal development of bone.

Osteogenesis   Formation or development of bone.

Osteoporosis   Significant decrease in bone mass with increased porosity and increased tendency to fracture.

Ovicide   Substance intended to kill eggs.

Paraesthesia   Abnormal sensation, as burning or prickling.
Para-occupational exposure
  1. Exposure of a worker's family to substances carried from the workplace to the home.
  2. Exposure of visitors to substances in the workplace.
Parasympatholytic   Producing effects resembling those caused by interruption of the parasympathetic nerve; also called anticholinergic.

Parasympathomimetic   Producing effects resembling those caused by stimulation of the parasympathetic nervous system; also called cholinomimetic.

Parenteral dosage   Method of introducing substances into an organism avoiding the gastrointestinal tract (subcutaneously, intravenously, intramuscularly *etc.*).

Paresis   Slight or incomplete paralysis.

Passive smoking   Inhalation of sidestream smoke by people who do not smoke themselves. See RT **sidestream smoke**.

Percutaneous   Through the skin following application on the skin.

Perinatal   Relating to the period shortly before and after birth; from the 20th to the 29th week of gestation to 1 to 4 weeks after birth.

Peritoneal dialysis   Method of artificial detoxication in which a toxic substance from the body is transferred into liquid that is instilled into the peritoneum. Thus, the employment of the peritoneum surrounding the abdominal cavity as a dialysing membrane for the purpose of removing waste products or toxins accumulated as a result of renal failure.

Permissible exposure limit (PEL) Recommendation by US OSHA for time-weighted average concentration that must not be exceeded during any 8 hour work shift of a 40 h working week. RT **maximum allowable concentration, threshold limit value, time weighted average concentration (TWAC), exposure limit**.

Peroxisome Organelle, similar to a lysosome, characterized by its content of catalase (EC 1.11.1.6), peroxidase (EC 1.11.1.7), and other oxidative enzymes.

Personal sampler Compact, portable instrument for individual air sampling, measuring, or both, the content of a harmful substance in the respiration zone of a working person. SN **individual monitor**.

Pest Organism that may harm public health, that attacks food and other materials essential to mankind, or otherwise affects human beings adversely.

Pesticide Strictly a substance intended to kill pests: in common usage, any substance used for controlling, preventing, or destroying animal, microbiological, or plant pests. NT **fungicide, herbicide, insecticide**.

Pesticide residue Pesticide residue is any substance or mixture of substances in food for man or animals resulting from the use of a pesticide and includes any specified derivatives, such as degradation and conversion products, metabolites, reaction products, and impurities considered to be of toxicological significance.

Phagocytosis Engulfing and digestion of micro-organisms, other cells, and foreign particles by cells such as phagocytes. RT **macrophage**.

Pharmacodynamics Process of interaction of pharmacologically active substances with target sites, and the biochemical and physiological consequences leading to therapeutic or adverse effects. RT **adverse effect, target, toxicodynamics**.

Pharmacogenetics Study of the influence of hereditary factors on the effects of drugs on individual organisms. PS **toxicogenetics**. RT **ecogenetics, polymorphism**.

Pharmacokinetics Process of the uptake of drugs by the body, the biotransformation they undergo, the distribution of the drugs and their metabolites in the tissues, and the elimination of the drugs and their metabolites from the body. Both the amounts and the concentrations of the drugs and their metabolites are studied. The term has essentially the same meaning as toxicokinetics, but the latter term should be restricted to the study of substances other than drugs. BT **chemobiokinetics**. PS **toxicokinetics**. RT **biotransformation**.

Pharynx Throat: the part of the digestive tract between the oesophagus, which is below, and the mouth and nasal cavities that are above and in front.

Phase 1 reaction (of biotransformation) Enzymic modification of a substance by oxidation, reduction, hydrolysis, hydration, dehydrochlorination, or other reactions catalysed by enzymes of the cytosol, of the endoplasmic reticulum (microsomal enzymes), or of other cell organelles. BT **biotransformation**. RT **cytochrome P420, cytochrome P448, cytochrome P450, microsome, phase 2 reaction, phase 3 reaction**.

Phase 2 reaction (of biotransformation) Binding of a substance or its metabolites from a phase 1 reaction, with endogenous molecules (conjugation), making more water-soluble derivatives that may be excreted in the urine or bile. BT **biotransformation**. RT **conjugate, phase 1 reaction, phase 3 reaction**.

Phase 3 reaction (of biotransformation) Further metabolism of conjugated metabolites produced by phase 2 reactions: this may result in the production of toxic derivatives. BT **biotransformation**. RT **conjugate, phase 1 reaction, phase 2 reaction**.

Phenotype The observable structural and functional characteristics of an organism determined by its genotype and modulated by its environment. RT **genotype**.

Pheromone Substance used in olfactory communication between organisms of the same species eliciting a change in sexual or social behaviour. SN **ectohormone, feromone**.

Photo-irritation Inflammation of the skin caused by exposure to light, especially that due to metabolites formed in the skin by photolysis. RT **photosensitization, phototoxicity**.

Photo-oxidant Substance able to cause oxidation when exposed to light of the appropriate wavelength.

Photophobia Abnormal visual intolerance of light.

Photosensitization Allergic reaction due to a metabolite formed by the influence of light.

Phototoxicity   Adverse effects produced by exposure to light energy, especially those produced in the skin.

Phytotoxic   Poisonous to plants; inhibiting plant growth.

Piscicide   Substance intended to kill fish.

Plasma
1. Fluid component of blood in which the blood cells and platelets are suspended. SN **blood plasma**.
2. Fluid component of semen produced by the accessory glands, the seminal vesicles, the prostate, and the bulbo-urethral glands.
3. Cell substance outside the nucleus. SN **cytoplasm**.

Plasma half-life   See SN **elimination half-life**

Plasmapheresis   Removal of blood from the body and centrifuging to obtain plasma and packed red blood cells: the blood cells are resuspended in a physiologically compatible solution (usually type-specific, fresh, frozen plasma or albumin) and returned to the donor or injected into a patient who requires blood cells rather than whole blood.

Plasmid   Autonomous, self-replicating extra-chromosomal, circular DNA molecule.

Pleura   Lining of the lung.

Ploidy   Term indicating the number of sets of chromosomes present in an organism. RT **diploid, haploid**.

Plumbism   Chronic poisoning caused by absorption of lead or lead salts. SN **saturnism**.

Pneumoconiosis   Usually fibrosis of the lungs that develops owing to (prolonged) inhalation of inorganic or organic dusts. Cause-specific types of pneumoconiosis:
1. **anthracosis**   From coal dust
2. **asbestosis**   From asbestos dust
3. **byssinosis**   From cotton dust
4. **siderosis**   From iron dust
5. **silicosis**   From silica dust
6. **stannosis**   From tin dust

Pneumonitis   Inflammation of the lung.

Po   *Per os* Latin for 'by mouth'.

Point mutation Reaction that changes a single base pair in DNA.

Point source Single emission source in a defined location. RT **area source**.

Poison Substance that, taken into or formed within the organism, impairs the health of the organism and may kill it. SN **toxic substance**.

Poison-bearing Containing a poison.

Poisoning Morbid condition produced by a poison. SN **intoxication**.

Pollutant Any undesirable solid, liquid, or gaseous matter in a solid, liquid, or gaseous environmental medium: 'undesirability' is often concentration-dependent, low concentrations of most substances being tolerable or even essential in many cases. For the meaning of 'undesirable' in air pollution contexts, see 'pollution'. A primary pollutant is one emitted into the atmosphere, water, sediments, or soil from an identifiable source. A secondary pollutant is a pollutant formed by chemical reaction in the atmosphere, water, sediments, or soil. PS **contaminant**. RT **pollution, secondary pollutant**.

Pollution Introduction of pollutants into a solid, liquid, or gaseous environmental medium, the presence of pollutants in a solid, liquid, or gaseous environmental medium, or any undesirable modification of the composition of a solid, liquid, or gaseous environmental medium . In the context of air pollution, an undesirable modification is one that has injurious or deleterious effects. RT **contaminant, pollutant**.

Polyclonal antibody Antibody produced by a number of different cell types. BT **antibody**. RT **monoclonal antibody**.

Polydipsia Chronic excessive thirst.

Polymorphism (polymorphia) in metabolism Inter-individual variations in metabolism of endo- and exogenous compounds due to genetic influences, leading to enhanced side effects or toxicity of drugs (*e.g.* poor versus fast metabolizers) or to different clinical effects (metabolism of steroid hormones). RT **ecogenetics, pharmacogenetics, toxicogenetics**.

Polyuria Excessive production and discharge of urine.

Population critical concentration (PCC) Concentration of a substance in the critical organ at which a specified percentage of the exposed population has reached the individual critical organ concentration (percentage indicated by PCC-10 for 10%, PCC-50 for 50% *etc.*; similar to the use of the term LD_{50}).

Population effect Absolute number or incidence rate of cases occurring in a group of people.

Population risk See SN **societal risk**

Porphyria Disturbance of porphyrin metabolism characterized by increased formation, accumulation, and excretion of porphyrins and their precursors.

Posology Study of dose in relation to the physiological factors that may influence response such as age of the exposed organisms.

Potency Expression of chemical or medicinal activity of a substance as compared with a given or implied standard or reference.

Potentiation Dependent action in which a substance or physical agent at a concentration or dose that does not itself have an adverse effect enhances the harm done by another substance or physical agent. RT **additive effect, antagonism, synergism**.

Practical certainty (of safety) Numerically specified low risk of exposure to a potentially toxic substance (*e.g.* 1 in 10^6) or socially acceptable low risk of adverse effects from such an exposure applied to decision making in regard to chemical safety. RT **risk, safety**.

Precordial Pertaining to the region over the heart and lower thorax.

Precursor Substance from which another, usually more biologically active, substance is formed.

Pre-neoplastic Before the formation of a tumour.

Prevalence Number of instances of existing cases of a given disease or other condition in a given population at a designated time; sometimes used to mean prevalence rate. When used without qualification, the term usually refers to the situation at a specified point in time (point prevalence). RT **incidence**.

Prevalence rate (ratio) Total number of individuals who have an attribute or disease at a particular time (or during a particular period) divided by the population at risk of having the attribute or disease at this point in time or midway through the period. RT **population at risk**.

Primary pollutant See BT **pollutant**

Probit Probability unit obtained by adding 5 to the normal deviates of a standardized normal distribution of results from a dose–response study: addition of 5 removes the complication of handling negative values. A plot of probit against the logarithm of dose or concentration gives a linear plot if the distribution of response is a logarithmic normal one. Estimates of the LD_{50} and ED_{50} (or LC_{50} and EC_{50}) can be obtained from this plot.

Procarcinogen   Substance that has to be metabolized before it can induce malignant tumours.

Prokaryote   Unicellular organism, characterized by the absence of a membrane-enclosed nucleus. Prokaryotes include bacteria, blue-green algae, and mycoplasmas. RT **eukaryote**.

Promoter (in oncology)   Agent that induces cancer when administered to an animal or human being who has been exposed to a cancer initiator. RT **initiator**.

Proportional mortality rate (ratio) (PMR)   Number of deaths from a given cause in a specified time period, per 100 or per 1000 total deaths in the same time period: can give rise to misleading conclusions if used to compare mortality experience of populations with different causes of death.

Prospective cohort study   See BT **cohort study**

Proteinuria   Excretion of excessive amounts of protein (derived from blood plasma or kidney tubules) in the urine.

Pseudoadaptation   Apparent adaptation of an organism to changing conditions of the environment (especially chemical) associated with stresses in biochemical systems that exceed the limits of normal (homeostatic) mechanisms: essentially there is a temporary, concealed pathology that later on can be manifested in the form of explicit pathological changes sometimes referred to as 'decompensation'. RT **compensation**.

Psychosis   Any major mental disorder characterized by derangement of the personality and loss of contact with reality.

Psychotropic   Exerting an effect upon the mind; capable of modifying mental activity.

Pulmonary   Pertaining to the lungs.

Purgative   See SN **cathartic, laxative**

Pyrexia   Condition in which the temperature of a human being or mammal is above normal.

Pyrogen   Any substance that produces fever.

Quantal effect   Condition that can be expressed only as 'occurring' or 'not occurring', such as death or occurrence of a tumour. AN **graded effect**. RT **stochastic effect**. SN **all-or-none effect**.

Quantitative structure–activity relationship (QSAR)   Quantitative association between the physicochemical properties of a substance and/or the properties of its molecular substructures and its biological properties, including its toxicity. RT **surrogate**.

Rate difference (RD)   Absolute difference between two rates, *e.g.* the difference in incidence rate between a population group exposed to a causal factor and a population group not exposed to the factor: in comparisons of exposed and unexposed groups, the term 'excess rate' may be used as a synonym for rate difference.

Rate ratio (RR)   In epidemiology, the value obtained by dividing the rate in an exposed population by the rate in an unexposed population.

Ratticide   Substance intended to kill rats. RT **rodenticide**.

Reference concentration   Term used for an estimate of air exposure concentration to the human population (including sensitive subgroups) that is likely to be without appreciable risk of deleterious effects during a lifetime. RT **acceptable daily intake**. BT **dose**.

Reference distribution   Statistical distribution of reference values.

Reference dose   Term used for an estimate (with uncertainty spanning perhaps an order of magnitude) of a daily exposure to the human population (including sensitive subgroups) that is likely to be without appreciable risk of deleterious effects during a lifetime. RT **acceptable daily intake**. BT **dose**.

Reference group   See SN **reference sample group**

Reference individual   Person selected with the use of defined criteria for comparative purposes in a clinical study.

Reference interval   Area between and including two reference limits, *e.g.* the percentiles 2.5 and 97.5.

Reference limit   Boundary value defined so that a stated fraction of the reference values is less than or exceeds that boundary value with a stated probability.

Reference material Substance for which one or more properties are sufficiently well established to be used for the calibration of an apparatus, the assessment of a measurement method, or for assigning values to other substances. SN **calibration material, standard material**.

Reference population Group of all reference individuals used to establish criteria against which a population that is being studied can be compared.

Reference sample group Selected reference individuals, statistically adequate numerically to represent the reference population.

Reference value According to the International Federation of Clinical Chemistry (IFCC), measured value of a property in a reference individual or sample from a reference individual. .

Relative odds See SN **odds ratio**

Relative risk
1. Ratio of the risk of disease or death among the exposed to that among the unexposed. SN **risk ratio**.
2. Ratio of the cumulative incidence rate in the exposed to the cumulative incidence rate in the unexposed; the cumulative incidence ratio.

Renal Pertaining to the kidneys.

Reproductive toxicant Substance or preparation that produces non-heritable harmful effects on the progeny and/or an impairment of male and female reproductive function or capacity. RT **teratogen**.

Reproductive toxicology Study of the adverse effects of substances on the embryo, fetus, neonate, and prepubertal mammal and the adult reproductive and neuro-endocrine systems. RT **embryo, fetus, neonate**.

Response
1. Proportion of an exposed population with a defined effect or the proportion of a group of individuals that demonstrate a defined effect in a given time at a given dose rate. RT **dose–response relationship**.
2. Reaction of an organism or part of an organism (such as a muscle) to a stimulus.

Retrospective study Research design used to test aetiological hypotheses in which inferences about exposure to the putative causal factor(s) are derived from data relating to characteristics of the persons or organisms under study or to events or experiences in their past: the essential feature is that some of the persons under study have the disease or other outcome condition of interest, and their characteristics and past experiences are compared with those of other, unaffected persons. Persons who differ in the severity of the disease may also be compared. RT **case control study**.

Reverse transcription Process by which an RNA molecule is used as a template to make a single-stranded DNA copy.

Rhabdomyolysis Acute, fulminating, potentially lethal disease of skeletal muscle that causes disintegration of striated muscle fibres as evidenced by myoglobin in the blood and urine.

Rhinitis Inflammation of the nasal mucosa.

Risk
1. Possibility that a harmful event (death, injury, or loss) arising from exposure to a chemical or physical agent may occur under specific conditions.
2. Expected frequency of occurrence of a harmful event (death, injury, or loss) arising from exposure to a chemical or physical agent under specific conditions.

NT **acceptable risk, excess lifetime risk, extra risk, societal risk**. RT **hazard**.

Risk acceptance Decision that the risk associated with a given chemical exposure or an event leading to such exposure is low enough to be tolerated in order to gain associated benefits. RT **acceptable risk**.

Risk assessment Identification and quantification of the risk resulting from a specific use or occurrence of a chemical or physical agent, taking into account possible harmful effects on individual people or society of using the chemical or physical agent in the amount and manner proposed and all the possible routes of exposure. Quantification ideally requires the establishment of dose–effect and dose–response relationships in likely target individuals and populations. RT **exposure assessment, hazard identification, risk characterization, risk estimation, risk evaluation, risk identification, risk management, risk perception**.

Risk aversion Term used to describe the tendency of an individual person to avoid risk.

Risk characterization Outcome of hazard identification and risk estimation applied to a specific use of a substance or occurrence of an environmental health hazard: the assessment requires quantitative data on the exposure of organisms or people at risk in the specific situation. The end product is a quantitative statement about the proportion of organisms or people affected in a target population. RT **hazard identification, risk estimation.**

Risk communication Interpretation and communication of risk assessments in terms that are comprehensible to the general public or to others without specialist knowledge.

Risk *de minimis* Risk that is negligible and too small to be of societal concern (usually assumed to be a probability below 10^{-5} or 10^{-6}); can also mean 'virtually safe'. In the US, this is a legal term used to mean 'negligible risk to the individual'. SN **negligible risk**.

Risk estimation Assessment, with or without mathematical modelling, of the probability and nature of effects of exposure to a substance based on quantification of dose–effect and dose–response relationships for that substance and the population(s) and environmental components likely to be exposed and on assessment of the levels of potential exposure of people, organisms, and environment at risk. RT **risk evaluation**.

Risk evaluation Establishment of a qualitative or quantitative relationship between risks and benefits, involving the complex process of determining the significance of the identified hazards and estimated risks to those organisms or people concerned with or affected by them. RT **exposure evaluation, hazard identification, risk assessment, risk characterization, risk estimation, risk identification, risk perception**.

Risk identification Recognition of a potential hazard and definition of the factors required to assess the probability of exposure of organisms or people to that hazard and of harm resulting from such exposure.

Risk indicator See SN **risk marker**

Risk management Decision-making process involving considerations of political, social, economic, and engineering factors with relevant risk assessments relating to a potential hazard so as to develop, analyse, and compare regulatory options and to select the optimal regulatory response for safety from that hazard. Essentially risk management is the combination of three steps: risk evaluation; emission and exposure control; risk monitoring. RT **emission and exposure control, risk evaluation, risk monitoring**.

Risk marker Attribute that is associated with an increased probability of occurrence of a disease or other specified outcome and that can be used as an indicator of this increased risk: not necessarily a causal or pathogenic factor. SN **risk indicator**.

Risk monitoring Process of following up the decisions and actions within risk management in order to check whether the aims of reduced exposure and risk are achieved. BT **monitoring**. RT **risk management**.

Risk perception Subjective perception of the gravity or importance of the risk based on a person's knowledge of different risks and the moral, economic, and political judgement of their implications. RT **risk evaluation**.

Risk phrases Word groups identifying potential health or environmental hazards required under CPL Directives (European Community); may be incorporated into Safety Data Sheets.

Risk ratio Value obtained by dividing the probability of occurrence of a specific effect in one group by the probability of occurrence of the same effect in another group, or the value obtained by dividing the probability of occurrence of one potentially hazardous event by the probability of occurrence of another. Calculation of such ratios is used in choosing between options in risk management. RT **risk management**.

Risk-specific dose Amount of exposure corresponding to a specified level of risk.

Rodenticide Substance intended to kill rodents.

Safety Reciprocal of risk: practical certainty that injury will *not* result from a hazard under defined conditions.
1. Safety of a drug or other substance in the context of human health: the extent to which a substance may be used in the amount necessary for the intended purpose with a minimum risk of adverse health effects.
2. Safety (toxicological): The high probability that injury will not result from exposure to a substance under defined conditions of quantity and manner of use, ideally controlled to minimize exposure. RT **practical certainty**, **risk**.

Safety factor See SN **uncertainty factor**

Sarcoma Malignant tumour arising in a connective tissue and composed primarily of anaplastic cells resembling supportive tissue.

Saturnism Intoxication caused by lead. SN **plumbism**.

Sclerosis Hardening of an organ or tissue, especially that due to excessive growth of fibrous tissue.

Screening
1. Carrying out of a test or tests, examination(s), or procedure(s) in order to expose undetected abnormalities, unrecognized (incipient) diseases, or defects: examples are mass X-rays and cervical smears.
2. Pharmacological or toxicological screening consists of a specified set of

procedures to which a series of compounds is subjected to characterize pharmacological and toxicological properties and to establish dose–effect and dose–response relationships.

Screening level   Decision limit or cut-off point at which a screening test is regarded as positive.

Secondary metabolite   Product of biochemical processes other than the normal metabolic pathways, mostly produced in micro-organisms or plants after the phase of active growth and under conditions of nutrient deficiency.

Secondary pollutant   See BT **pollutant**

Secondhand smoke   See SN **sidestream smoke**

Secretion
  1. Process by which a substance such as a hormone or enzyme produced in a cell is passed through a plasma membrane to the outside, *e.g.* the intestinal lumen or the blood (internal secretion).
  2. Solid, liquid, or gaseous material passed from the inside of a cell through a plasma membrane to the outside as a result of cell activity.

Sedative   Substance that exerts a soothing or tranquillizing effect. RT **anaesthetic, narcotic**.

Semichronic   See SN **subchronic**

Sensibilization   See SN **sensitization**

Sensitivity (of a screening test)   Extent (usually expressed as a percentage) to which a method gives results that are free from false negatives; the fewer the false negatives, the greater the sensitivity. Quantitatively, sensitivity is the proportion of truly diseased persons in the screened population who are identified as diseased by the screening test. RT **specificity (of a screening test)**.

Sensitization   Immune process whereby individuals become hypersensitive to substances, pollen, dandruff, or other agents that make them develop a potentially harmful allergy when they are subsequently exposed to the sensitizing material (allergen). RT **allergy, hypersensitivity**.

Sensory effect level
  1. Intensity, where the detection threshold level is defined as the lower limit of the perceived intensity range (by convention the lowest concentration that can be detected in 50% of the cases in which it is present).
  2. Quality, where the recognition threshold level is defined as the lowest

concentration at which the sensory effect can be recognized correctly in 50% of the cases.

3. Acceptability and annoyance, where the nuisance threshold level is defined as the concentration at which not more than a small proportion of the population (less than 5%) experiences annoyance for a small part of the time (less than 2%). Since annoyance will be influenced by a number of factors, a nuisance threshold level cannot be set on the basis of concentration alone. RT **nuisance threshold**.

Serum
1. Watery proteinaceous portion of the blood that remains after clotting. SN **blood serum**.
2. Clear watery fluid especially that moistening the surface of serous membranes or that exuded through inflammation of any of these membranes.

Short-term effect   See SN **acute effect**

Short-term exposure limit (STEL)   As used by the US NIOSH (National Institute of Occupational Safety and Health), unless noted otherwise, the 15 minute time-weighted average exposure that should not be exceeded at any time during a workday.

Side effect   Action of a drug other than that desired for beneficial pharmacological effect.

Siderosis
1. Pneumoconiosis resulting from the inhalation of iron dust. BT **pneumoconiosis**.
2. Excess of iron in the urine, blood, or tissues, characterized by haemosiderin granules in urine and iron deposits in tissues.

Sidestream smoke   Cloud of small particles and gases that is given off from the end of a burning tobacco product (cigarette, pipe, cigar) between puffs and is not directly inhaled by the smoker; the smoke that gives rise to passive inhalation on the part of bystanders. SN **secondhand smoke**. RT **mainstream smoke**.

Sign   Objective evidence of a disease, deformity, or an effect induced by an agent, perceptible to an examining physician.

Silicosis   Pneumoconiosis resulting from inhalation of silica dust. BT **pneumoconiosis**.

Sink   In environmental chemistry, an area or part of the environment in which, or a process by which, one or more pollutants is removed from the medium in which it is dispersed *e.g.* moist ground acts as a sink for sulfur dioxide in the air.

Sister chromatid exchange (SCE) Reciprocal exchange of chromatin between two replicated chromosomes that remain attached to each other until anaphase of mitosis; used as a measure of mutagenicity of substances that produce this effect. RT **mitosis**.

Skeletal fluorosis Osteosclerosis due to fluoride.

Slimicide Substance intended to kill slime-producing organisms (used on paper stock, water cooling systems, paving stones *etc.*).

Slope factor Value, in inverse concentration or dose units, derived from the slope of a dose–response curve; in practice, limited to carcinogenic effects with the curve assumed to be linear at low concentrations or doses. The product of the slope factor and the exposure is taken to reflect the probability of producing the related effect. RT **concentration–effect curve, concentration–response curve, dose, dose–effect curve, dose–response curve**.

Societal risk Total probability of harm to a human population including also the probability of adverse health effects to descendants and the probability of disruption resulting from loss of services such as industrial plant or loss of material goods and electricity.

Solvent abuse Deliberate inhalation (or drinking) of volatile solvents, in order to become intoxicated. SN **'solvent sniffing'**. NT **'glue sniffing'**.

'Solvent sniffing' See SN **solvent abuse**. NT **'glue sniffing'**

Somatic
1. Pertaining to the body as opposed to the mind.
2. Pertaining to non-reproductive cells or tissues.
3. Pertaining to the framework of the body as opposed to the viscera.

Soporific Substance producing sleep. RT **anaesthetic, narcotic, sedative**.

Speciation Determination of the exact chemical form or compound in which an element occurs in a sample, for instance, determination of whether arsenic occurs in the form of trivalent or pentavalent ions or as part of an organic molecule, and the quantitative distribution of the different chemical forms that may coexist.

Species
1. In biological systematics, group of organisms of common ancestry that are able to reproduce only among themselves and that are usually geographically distinct.
2. See NT **chemical species**.

Species differences in sensitivity Quantitative or qualitative differences of response to the action(s) of a potentially toxic substance on various species of living organisms. RT **species-specific sensitivity**.

Species-specific sensitivity Quantitative and qualitative features of response to the action(s) of a potentially toxic substance that are characteristic for particular species of living organism. RT **species differences in sensitivity**.

Specific death rate Death rate computed for a subpopulation of individual organisms or people having a specified characteristic or attribute, and named accordingly (*e.g.* age-specific death rate: the number of deaths of persons of a specified age during a given period of time, divided by the total number of persons of that age in the population during that time).

Specificity (of a screening test) Proportion of truly non-diseased persons who are identified by the screening test.

Specific pathogen-free (SPF) Describing an animal removed from its mother under sterile conditions just prior to term and subsequently reared and kept under sterile conditions. RT **germ-free animal**.

Spreader Agent used in some pesticide formulations to extend the even disposition of the active ingredient.

Stability half-life (half-time) Time required for the amount of a substance in a formulation to decrease, for any reason, by one-half (50%).

Standard(ized) mortality (morbidity) ratio (SMR) Ratio of the number of events observed in the study group or population to the number of deaths expected if the study population had the same specific rates as the standard population (multiplied by 100).

Stannosis Pneumoconiosis resulting from inhalation of tin dust.

Stochastic Of, pertaining to, or arising from chance and therefore involving probability and obeying the laws of probability.

Stochastic effect Consequence for which the probability of occurrence depends on the absorbed dose: hereditary effects and cancer induced by radiation are considered to be stochastic effects. The term 'stochastic' indicates that the occurrence of effects so named would be random. This means that, even for an individual, there is no threshold of dose below which the effect will not appear, and the chance of experiencing the effect increases with increasing dose. RT **all-or-none effect, quantal effect**.

Stratification (in epidemiology) Process of or result of separating a sample into several subsamples according to specified criteria such as age groups, socio-economic status, *etc.*

Stratified sample   Subset of a population selected according to some impor-tant characteristic. RT **stratification**.

Structure–activity relationship (SAR)   Association between the physico-chemical properties of a substance and/or the properties of its molecular substructures and its biological properties, including its toxicity. PS **quantitative structure–activity relation (QSAR)**.

Subacute   Term used to describe a form of repeated exposure or administra-tion usually occurring over about 21 days, not long enough to be called 'long-term' or 'chronic'. PS **subchronic**. RT **subacute effect, subchronic effect, subchronic toxicity, subchronic toxicity test**.

Subacute (sometimes called subchronic) effect   Biological change re-sulting from multiple or continuous exposures usually occurring over about 21 days. Sometimes the term is used synonymously with subchronic effect and care should be taken to check the usage for any particular case. PS **subchronic effect**. RT **subchronic toxicity, subchronic toxicity test**.

Subchronic   Related to repeated dose exposure over a short period, usually about 10% of the life span. An imprecise term used to describe exposures of intermediate duration. PS **subacute**. RT **subacute effect, subchronic effect, subchronic toxicity, subchronic toxicity test**. SN **semichronic**.

Subchronic (sometimes called subacute) effect   Biological change re-sulting from an environmental alteration lasting about 10% of the lifetime of the test organism. In practice with experimental animals, such an effect is usually identified as resulting from multiple or continuous exposures occurring over 3 months (90 days). Sometimes a subchronic effect is distinguished from a subacute effect on the basis of its lasting for a much longer time. PS **subacute effect**. RT **subchronic toxicity, subchronic toxicity test**.

Subchronic toxicity
1. Adverse effects resulting from repeated dosage or exposure to a substance over a short period, usually about 10% of the life span.
2. The capacity to produce adverse effects following subchronic exposure. RT **subacute, subchronic, subchronic effect, subchronic toxicity test**.

Subchronic (sometimes called subacute) toxicity test   Animal experi-ment serving to study the effects produced by the test material when administered in repeated doses (or continually in food, drinking water, air) over a period of up to about 90 days. SN **semichronic toxicity test**.

Subclinical effect Biological change following exposure to an agent known to cause disease either before symptoms of the disease occur or when they are absent.

Subthreshold dose Amount of a substance that is lower than the amount which must be administered to produce the lowest detectable adverse effect.

Sudorific Substance that causes sweating.

Suggested no adverse response level (SNARL) Maximum dose or concentration that on current understanding is likely to be tolerated by an exposed organism without producing any harm.

Super-threshold dose See PS **toxic dose**

Surrogate Relatively well studied toxicant whose properties are assumed to apply to an entire chemically and toxicologically related class; *e.g.* benzo[*a*]pyrene data may be used as toxicologically equivalent to that for all carcinogenic polynuclear aromatic hydrocarbons. RT **quantitative structure–activity relationship**.

Surveillance Ongoing scrutiny, generally using methods distinguished by their practicability and uniformity, and frequently by their rapidity, rather than by complete accuracy. Its main purpose is to detect changes in trend or distribution in order to initiate investigative or control measures.

Sympatholytic
1. adj. Blocking transmission of impulses from the adrenergic (sympathetic) postganglionic fibres to effector organs or tissues.
2. n. Agent that blocks transmission of impulses from the adrenergic (sympathetic) postganglionic fibres to effector organs or tissues. SN **antiadrenergic**.

Sympathomimetic
1. adj. Producing effects resembling those of impulses transmitted by the postganglionic fibres of the sympathetic nervous system.
2. n. Agent that produces effects resembling those of impulses transmitted by the postganglionic fibres of the sympathetic nervous system. SN **adrenergic**.

Symptom Any subjective evidence of a disease or an effect induced by a substance as perceived by the affected subject.

Synapse   Functional junction between two neurones, where a nerve impulse is transmitted from one neurone to another.

Synaptic transmission   See RT **synapse**

Syndrome   Set of signs and symptoms occurring together and often characterizing a particular disease-like state.

Synergism   Pharmacological or toxicological interaction in which the combined biological effect of two or more substances is greater than expected on the basis of the simple summation of the toxicity of each of the individual substances.

Synergistic effect   Biological effect following exposure simultaneously to two or more substances that is greater than the simple sum of the effects that occur following exposure to the substances separately. RT **additive effect**, **antagonism**, **potentiation**.

Systemic   Relating to the body as a whole.

Systemic effect   Consequence that is of either a generalized nature or that occurs at a site distant from the point of entry of a substance: a systemic effect requires absorption and distribution of the substance in the body.

Tachy-   Prefix meaning rapid, as in tachycardia and tachypnoea.

Tachycardia   Abnormally fast heartbeat. AN **bradycardia**.

Tachypnoea   Abnormally fast breathing. AN **bradypnoea**.

Taeniacide   Substance intended to kill tapeworms.

Target (biological)   Any organism, organ, tissue, cell, or cell constituent that is subject to the action of a pollutant or other chemical, physical, or biological agent. RT **receptor**.

Target organ(s)   Organ(s) in which the toxic injury manifests itself in terms of dysfunction or overt disease. RT **receptor**.

Target population (epidemiology)
  1. Collection of individuals, items, measurements, *etc.* about which we want to make inferences: the term is sometimes used to indicate the population from which a sample is drawn and sometimes to denote any reference population about which inferences are required.
  2. Group of persons for whom an intervention is planned.

T-cell   See SN **T-lymphocyte**

Teratogen   Agent that, when administered prenatally (to the mother), induces permanent structural malformations or defects in the offspring.

Teratogenicity   Potential to cause or the production of structural malformations or defects in offspring. RT **developmental toxicity**, **embryotoxicity**.

Tetanic   Pertaining to tetanus, characterized by tonic muscle spasm.

Therapeutic index   Ratio between toxic and therapeutic doses (the higher the ratio, the greater the safety of the therapeutic dose).

Threshold   Dose or exposure concentration below which an effect is not expected.

Threshold limit value (TLV)   Concentration in air of a substance to which it is believed that most workers can be exposed daily without adverse effect (the threshold between safe and dangerous concentrations). These values are established (and revised annually) by the American Conference of Governmental Industrial Hygienists) and are time-weighted concentrations for a 7 or 8 hour workday and a 40 hour workweek. For most substances the value may be exceeded, to a certain extent, provided there are compensatory periods of exposure below the value during the workday (or in some cases the week). For a few substances (mainly those that produce a rapid response) the limit is given as a ceiling concentration (maximum permissible concentration, designated by 'C') that should never be exceeded.

Thrombocytopenia   Decrease in the number of blood platelets (thrombocytes).

Tidal volume   Quantity of air or test gas that is inhaled and exhaled during one respiratory cycle.

Time-weighted average exposure (TWAE) or concentration (TWAC)   Concentration in the exposure medium at each measured time interval, multiplied by that time interval, and divided by the total time of observation: for occupational exposure a working shift of eight hours is commonly used as the averaging time.

Tinnitus   Continual noise in the ears, such as ringing, buzzing, roaring, or clicking.

Tissue dose   Amount of a substance or physical agent (radiation) absorbed by a tissue.

T-lymphocyte Animal cell that possesses specific cell surface receptors through which it binds to foreign substances or organisms, or those that it identifies as foreign, and which initiates immune responses. RT **B-lymphocyte, immune response, lymphocyte**.

Tolerable daily intake (TDI) Regulatory value equivalent to the acceptable daily intake established by the European Commission Scientific Committee on Food. Unlike the ADI, the TDI is expressed in mg/person, assuming a body weight of 60 kg. TDI is normally used for food contaminants. RT **acceptable daily intake**.

Tolerable risk Probability of suffering disease or injury that can, for the time being, be tolerated, taking into account the associated benefits, and assuming that the risk is minimized by appropriate control procedures. PS **acceptable risk**.

Tolerance
1. Adaptive state characterized by diminished effects of a particular dose of a substance: the process leading to tolerance is called 'adaptation'.
2. In food toxicology, dose that an individual can tolerate without showing an effect.
3. Ability to experience exposure to potentially harmful amounts of a substance without showing an adverse effect.
4. Ability of an organism to survive in the presence of a toxic substance: increased tolerance may be acquired by adaptation to constant exposure.
5. In immunology, state of specific immunological unresponsiveness.

Tonic
1. Characterized by tension, especially muscular tension.
2. Medical preparation that increases or restores normal muscular tension.

Topical Pertaining to a particular area, as in a topical effect that involves only the area to which the causative substance has been applied.

Toxic Able to cause injury to living organisms as a result of physicochemical interaction.

Toxicant See SN **toxic substance**

Toxic chemical See SN **toxic substance**

Toxic dose Amount of a substance that produces intoxication without lethal outcome. SN **super-threshold dose**.

Toxicity
1. Capacity to cause injury to a living organism, defined with reference to the

quantity of substance administered or absorbed, the way in which the substance is administered (inhalation, ingestion, topical application, injection) and distributed in time (single or repeated doses), the type and severity of injury, the time needed to produce the injury, the nature of the organism(s) affected, and other relevant conditions.

2. Adverse effects of a substance on a living organism, defined with reference to the quantity of substance administered or absorbed, the way in which the substance is administered (inhalation, ingestion, topical application, injection) and distributed in time (single or repeated doses), the type and severity of injury, the time needed to produce the injury, the nature of the organism(s) affected, and other relevant conditions.

3. Measure of incompatibility of a substance with life: this quantity may be expressed as the reciprocal of the absolute value of median lethal dose $(1/LD_{50})$ or concentration $(1/LC_{50})$. RT **acute toxicity**, **chronic toxicity**, **subacute toxicity**, **subchronic toxicity**.

Toxicity equivalency factor (TEF)   Factor used in risk assessment to estimate the toxicity of a complex mixture, most commonly a mixture of chlorinated dibenzo-*p*-dioxins, furans, and biphenyls: TEF is usually based on relative toxicity to 2,3,7,8-tetrachlorodibenzo-*p*-dioxin (TEF = 1).

Toxicity equivalent (TEQ)   Contribution of a specified component (or components) to the toxicity of a mixture of related substances. The amount of substance (or substance concentration) of total toxicity equivalent is the sum of that for the components B, C ... N:

$$S\,n(TEQ) = n(TEQ)_B + n(TEQ)_C + \ldots n(TEQ)_N$$

Toxicity equivalent is most commonly used in relation to the reference toxicant 2,3,7,8-tetrachlorodibenzo-*p*-dioxin (2,3,7,8-TCDD) by means of the toxicity equivalency factor (TEF, f) which is 1 for the reference substance, hence:

$$S\,n(TEQ) = f_B n_B + f_C n_C + \ldots f_N n_N$$

Toxicity test   Experimental study of the adverse effects of exposure of a living organism to a substance for a defined duration under defined conditions. RT **acute toxicity test**, **carcinogenicity test**, **chronic toxicity test**, **subchronic toxicity test**.

Toxic material   See SN **toxic substance**

Toxicodynamics   Process of interaction of potentially toxic substances with target sites, and the biochemical and physiological consequences leading to adverse effects. RT **adverse effect**, **pharmacodynamics**, **target**.

Toxicogenetics   Study of the influence of hereditary factors on the effects of potentially toxic substances on individual organisms. RT **ecogenetics**, **pharmacogenetics, polymorphism**.

Toxicokinetics   Process of the uptake of potentially toxic substances by the body, the biotransformation they undergo, the distribution of the substances and their metabolites in the tissues, and the elimination of the substances and their metabolites from the body. Both the amounts and the concentrations of the substances and their metabolites are studied. The term has essentially the same meaning as pharmacokinetics, but the latter term should be restricted to the study of pharmaceutical substances. BT **chemobiokinetics**. RT **biotransformation, pharmacokinetics**.

Toxicological data sheet   Document that gives in a uniform manner data relating to the toxicology of a substance, its production and application, properties and methods of identification; the data sheet may also include recommendations on protective measures. PS **toxicological profile, toxicological dossier**.

Toxicology   Scientific discipline involving the study of the actual or potential danger presented by the harmful effects of substances (poisons) on living organisms and ecosystems, of the relationship of such harmful effects to exposure, and of the mechanisms of action, diagnosis, prevention, and treatment of intoxications. NT **chemical toxicology.**

Toxicometry   Term sometimes used to indicate a combination of investigative methods and techniques for making a quantitative assessment of toxicity and the hazards of potentially toxic substances.

Toxicophobia   Morbid dread of poisons. RT **chemophobia**.

Toxicophoric (toxophoric) group   Structural moiety that upon metabolic activation exerts toxic effects: the presence of a toxicophoric group indicates only potential and not necessarily actual toxicity of a drug or other substances. SN **toxogenic group**.

Toxicovigilance   Active process of identification, investigation, and evaluation of various toxic effects in the community with a view to taking measures to reduce or control exposure(s) involving the substance(s) that produce(s) these effects.

Toxic substance   Material causing injury to living organisms as a result of physicochemical interactions.   SN **chemical etiologic agent, poison, toxicant, toxic chemical, toxic material**.

Toxification   Metabolic conversion of a potentially toxic substance into a product that is more toxic.

Toxin Poisonous substance produced by a biological organism such as a microbe, animal, or plant. PS **venom**.

Toxinology Scientific discipline involving the study of the chemistry, biochemistry, pharmacology, and toxicology of toxins. RT **toxicology, toxin**.

Toxogenic group See SN **toxicophoric group**

Transcription Process by which the genetic information encoded in a linear sequence of nucleotides in one strand of DNA is copied into an exactly complementary sequence of RNA. RT **reverse transcription**.

Transformation
1. Alteration of a cell by incorporation of foreign genetic material and its subsequent expression in a new phenotype. RT **phenotype**.
2. Conversion of cells growing normally to a state of rapid division in culture resembling that of a tumour.
3. Chemical modification of substances in the environment.

Trophic level Amount of energy in terms of food that an organism needs: organisms not needing organic food, such as plants, are said to be on a low trophic level, whereas predator species needing food of high energy content are said to be on a high trophic level. The trophic level indicates the level of the organism in the food chain.

Tumorigenic Able to cause tumours.

Tumour
1. Any abnormal swelling or growth of tissue, whether benign or malignant.
2. An abnormal growth, in rate and structure, that arises from normal tissue, but serves no physiological function. SN **neoplasm**.

Tumour progression Sequence of changes by which a benign tumour develops from the initial lesion to a malignant stage.

Ulcer Defect, often associated with inflammation, occurring locally or at the surface of an organ or tissue owing to sloughing of necrotic tissue.

Uncertainty factor
1. In assay methodology, confidence interval or fiducial limit used to assess the probable precision of an estimate.
2. In toxicology, value used in extrapolation from experimental animals to man (assuming that man may be more sensitive) or from selected individuals to the general population:*e.g.* a value applied to the no observed effect level (NOEL) or no observed adverse effect level (NOAEL) to derive an acceptable daily intake or reference dose (RfD) (the NOEL or NOAEL is

divided by the value to calculate the acceptable daily intake or RfD). The value depends on the nature of the toxic effect, the size and type of population to be protected, and the quality of the toxicological information available. SN **assessment factor, modifying factor, safety factor**. RT **no observed effect level, no observed adverse effect-level, reference dose**.

Urticaria   Vascular reaction of the skin marked by the transient appearance of smooth, slightly elevated patches (wheals, hives) that are redder or paler than the surrounding skin and often attended by severe itching.

Vacuole   Membrane-bound cavity within a cell.

Vasoconstriction   Decrease of the calibre of the blood vessels leading to a decreased blood flow. AN **vasodilation**.

Vasodilation   Increase in the calibre of the blood vessels, leading to an increased blood flow. AN **vasoconstriction**.

Vehicle   Substance(s) used to formulate active ingredients for administration or use (general term for solvents, suspending agents *etc.*). RT **excipient**.

Venom   Animal toxin generally used for self-defence or predation and usually delivered by a bite or sting. PS **toxin**.

Ventilation
  1. Process of supplying a building or room with fresh air.
  2. Process of exchange of air between the ambient atmosphere and the lungs.
  3. In physiology, the amount of air inhaled per day.
  4. Oxygenation of blood.

Ventricular fibrillation   Irregular heartbeat characterized by unco-ordinated contractions of the ventricle.

Vermicide   Substance intended to kill worms.

Vermifuge   Substance that causes the expulsion of intestinal worms.

Vertigo   Dizziness: an illusion of movement as if the external world were revolving around an individual or as if the individual were revolving in space.

Vesicant
  1. adj. Producing blisters on the skin.
  2. n. Substance that causes blisters on the skin.

Vesicle
  1. Small sac or bladder containing fluid.

2. Blister-like elevation on the skin containing serous fluid.

Volume of distribution   Apparent (hypothetical) volume of fluid required to contain the total amount of a substance in the body at the same concentration as that present in the plasma assuming equilibrium has been attained.

Waste   Anything that is discarded deliberately or otherwise disposed of on the assumption that it is of no further use to the primary user.

Wasting syndrome   Disease marked by weight loss and atrophy of muscular and other connective tissues that is not directly related to a decrease in food and water consumption.

Withdrawal effect   Adverse event following withdrawal from a person or animal of a drug to which they have been chronically exposed or on which they have become dependent.

X-disease   Hyperkeratotic disease in cattle following exposure to chlorinated dibenzo-*p*-dioxins, naphthalenes, and related compounds.

Xenobiotic
1. Strictly, any substance interacting with an organism that is not a natural component of that organism. SN **exogenous substance, foreign substance or compound**.
2. Man-made compounds with chemical structures foreign to a given organism. SN **anthropogenic substance**.

Zoocide   Substance intended to kill animals.

Zygote
1. Cell, such as a fertilized egg, resulting from the fusion of two gametes.
2. Cell obtained as a result of complete or partial fusion of cells produced by meiosis.

Subject Index